BEYOND THE HORIZON, The History of Airborne Early Warning

Ian Shaw with Sérgio Santana

BEYOND THE HORIZON
The History of Airborne Early Warning

Ian Shaw with Sérgio Santana

Copyright © 2014 Harpia Publishing L.L.C. and Moran Publishing L.L.C. Joint Venture
2803 Sackett Street, Houston, TX 77098-1125, U.S.A.
aewc@harpia-publishing.com

All rights reserved.

No part of this publication may be copied, reproduced, stored electronically or transmitted
in any manner or in any form whatsoever without the written permission of the publisher.

Consulting and inspiration by Kerstin Berger
Artworks and drawings by Tom Cooper and James Lawrence
Maps by James Lawrence
Editorial by Thomas Newdick
Layout by Norbert Novak, www.media-n.at, Vienna
and James Lawrence, www.gingercake-creative.co.uk, United Kingdom

Harpia Publishing, L.L.C. is a member of

Printed at Grasl Druck & Neue Medien, Austria

ISBN 978-0-9854554-3-9

Contents

Introduction 6
Acknowledgements 8
Abbreviations 10
Chapter 1: First Steps, 1943–1963 15
Chapter 2: Going to War, 1964–1968 47
Chapter 3: Coming of Age, 1969–1980 63
Chapter 4: Crucial Years, 1980–1990 89
Chapter 5: The New Era, 1991–2001 101
Chapter 6: War on Terror, 2001 to today 127
Chapter 7: AEW&C Operators today 143
Appendix I: Technical specifications of AEW aircraft 185
   1   AgustaWestland AW101 Mk 112 186
   2   Avro Shackleton AEW.Mk 2 188
   3   Beriev A-50, Baghdad/Adnan and Xi'an KJ-2000 190
   4   Boeing 737 AEW&C 192
   5   Boeing E-3 Sentry 194
   6   Boeing E-767 196
   7   Boeing PB-1W 198
   8   Changhe Z-18Y Black Bat 200
   9   Douglas AD-3W, 4W and 5W Skyraider 202
  10   Embraer 145 AEW&C 204
  11   Fairey Gannet AEW.Mk 3 206
  12   Goodyear ZPG-2W/3W 208
  13   Grumman E-1 Tracer 210
  14   Grumman E-2 Hawkeye 212
  15   Grumman TBM-3W Avenger 216
  16   Gulfstream 550 CAEW 218
  17   IAI Condór 220
  18   Kamov Ka-31 222
  19   Lockheed Neptune AEW.Mk 1 224
  20   Lockheed P-3 AEW 226
  21   Lockheed WV and EC-121 Warning Star 228
  22   Saab 340 AEW&C and Saab 2000 AEW&C 230
  23   Shaanxi KJ-200, ZDK-03 and Y-8J 232
  24   Tupolev Tu-126 234
  25   Vickers Wellington Mk IC and Mk XIV 236
  26   Westland Sea King AEW.Mk 2/ASaC.Mk 7 & Sikorsky SH-3H(AEW) Sea King ... 238
Appendix II: Service Time Lines and Radar Types and Ranges 241
Bibliography 245
Index 251

Introduction

Very little has been published about airborne early warning (AEW) aircraft and their history. This is unusual considering the importance of these platforms for modern air forces, and the fact that they have repeatedly played a crucial role in recent conflicts. It is strange too, because the development of AEW aircraft since 1945 had been far from rapid and has experienced a few false starts and developmental disasters. Rapid technological development, miniaturisation, and multilateral collaboration between many smaller nations today have changed the AEW landscape such that it is no longer dominated by the major powers. Some of the 'smaller players' are meanwhile producing very effective and relatively cheap systems, and are more than willing to sell them.

The primary purpose of this book is to provide a complete and unbiased account of the early history and development of AEW aircraft and their systems, without baffling the reader with scientific and technical jargon. All nations that have been involved in AEW developments are covered comprehensively.

The start of this story can be traced back to the early days of World War II, when the United Kingdom installed a comprehensive network of ground-based radar stations along its coasts facing continental Europe. This radar chain proved essential during the Battle of Britain, in late summer 1940, by providing timely warning of incoming formations of German bombers and enabling the Royal Air Force (RAF) to intercept them from favourable positions. Germany knew about the radar and had been trying to develop equivalent systems, but did not realise the importance of the British radar network and thus never seriously attempted to disrupt its operation.

There was one major problem with ground-based radar stations: they could not see beyond the horizon, and radar towers could not be raised high enough above the ground to improve this significantly, and thus it did not take very long before airmen realised that they could fly 'below the radar horizon', to avoid early detection. The answer to this problem was found by installing radars on board aircraft, enabling them to 'see' beyond the horizon, to look down on hostile aircraft as they approached their targets.

While sounding simple, this idea proved very hard to realise, because contemporary radars were heavy, bulky, weak, and not suitable to being bumped around in an aircraft. Thus the race began to develop small, light, powerful, robust and reliable, airportable radar systems. Spurred by war, this goal was not only achieved, but also multifunctional systems began to emerge, capable of detecting enemy ships, for example.

The next issue to be solved was that of passing the resulting information to command centres and friendly assets best suited to deal with the threat. The airborne early warning and control (AEW&C) concept was born with the Wellington ACI, which reunited these capabilities in a single platform, as it was designed to detect targets *and* guide allied aircraft against them.

Essentially, an airborne early warning and control system is a complex, mission-specialised system, designed to detect enemy forces at long range and then direct friendly assets against them. First conceived in the United Kingdom during World War II, it was developed by the United States into the form in which it is known today.

Installed on airborne platforms, AEW&C systems rapidly proved able of surpassing geographical limits and evolved into a true 'force multiplier'. Through installation of sub-systems collecting electronic and signals intelligence (ELINT and SIGINT, respectively), and meanwhile often supported by unmanned aerial vehicles (UAVs), such aircraft are nowadays recognised as one of the core requirements of the 'net-centric warfare' concept, in which all the information collected by a nation's different intelligence and military assets is combined to improve the situational awareness of involved commanders and their 'war fighters' by several magnitudes. Modern AEW&C platforms do not only serve the purpose of early detection, but actually represent highly effective decision-making centres, from which commanders can effectively manage large and diverse military forces moving over a vast battle space, without the limitations and dangers faced by immobile ground-based stations.

This book has been compiled on the basis of a coherent, chronological coverage and inclusion of 'crew stories'. Realising that military developments are driven by needs within specific contexts, especially during an ongoing conflict, the authors developed a basic timeline, and then introduced specific aircraft types along it, reflecting on their development before going forward to describe the types' operational service. The most important conflicts in terms of the deployment of AEW&C platforms have been selected, and a study is provided of how the conflicts in question influenced the further development of such aircraft. We have excluded any 'what ifs' and types that have had only a limited early warning role, or that lack the 'control' capability, as they do not meet the established objective of this book.

Correspondingly, the first part of this book tells the story of AEW&C aircraft, their roles and development in all major conflicts since World War II. The second purpose of this book is to provide detailed, precise and above all accurate information covering related events from the early 1940s until today, many of which have been overlooked or omitted by other authors.

The second half of the book is devoted to the details, colour schemes and operators of individual aircraft types. Finally, the appendices list aircraft specifications, , types of AEW radar and their ranges. We hope we have managed to tell as complete a story as possible, the history and evolution of an essential war asset, whose overarching importance is often forgotten or underrated.

The third purpose of this book is to include first-person accounts. This book had to be written now, while the advent of AEW aircraft is still within living memory, just. Many of the air and ground crews who operated the very early aircraft described in this book have since passed away. It is vital that the survivors' recollections are recorded before they depart.

Wherever possible we have spoken to or written to the men and women who were there and took part. Obviously, many aspects of today's operations cannot be revealed in full, because of security concerns. Some authorities do not let their operational crews talk to the public, and attempts to contact them only resulted in stonewalling.

Input has therefore come from personal experiences, including one of the author's own service in the AEW community, the 'grapevine' and general knowledge of the subject, picked over years of working closely with the crews involved. Other information is drawn from official sources, often scattered in the public domain. The authors are confident that we will have to wait for another 40 or 50 years until the crews that operated in recent campaigns over Afghanistan, Libya or Syria will be free to talk about their own exploits.

Sérgio Santana and Ian Shaw, June 2014

Acknowledgements

Scores of people generously gave their time and effort in assisting in the compilation of this book, over the nine years it took to prepare. Ian Brown, the Assistant Curator of Aviation at the Museum of Flight was the very first historian to point me in the right direction. Sadly most of those involved in the early days of AEW have now left us, so we were very fortunate to receive a first-hand account of Operation Vapour from Ross Hamilton in Canada.

We tried to locate as many original AEW aircrews as possible and obtain their stories. We thank George E. Stewart, a flight engineer on PB-1Ws; Bert Udovin, a radar operator on the same type; and Pete Wasmund and Bill Green, who both flew WV-2s with the US Navy. Among their US Air Force colleagues, thanks in particular are due to Dean Boys, who was very generous in allowing us to use material from his website, and fellow 'Connie' crew members George Merryman and Don Born. Among those who described their missions in the E-1B off Yankee Station during the Vietnam War; former pilot Bob Braze cannot be thanked enough, and Steve Gorek, a Tracer radar man, gave us a genuine feel of what it was like to be 'down the back' in an early AEW aircraft. 'Wild Bill' Richards provided invaluable insights into Operation Desert Storm and his excerpts from the Airborne Early Warning Association website.

Contributors from the UK included Royal Navy Gannet AEW.Mk 3 pilot Ian Parkinson, and Royal Air Force Shackleton AEW.Mk 2 crew members Bill Howard and Geoff Cooper, and Ken Walne, who began on the same type before an illustrious career on the Nimrod AEW and the E-3D Sentry. Chris Gibson, an outstanding aviation author, provided help and guidance along the way. A special mention has been reserved for Lt Cdr Colin McGannity, the commanding officer of 849 Naval Air Squadron (NAS) at RNAS Culdrose, Lt Martin Pittock, Lt Simon Richards and also Lt Cdr James Hall, commanding officer of 854 NAS, for a special day of briefings on the Sea King ASaC.Mk 7, its history and their role in the fleet.

We received considerable help from Hellenic Air Force pilot Kriakos 'Kirk' Paloulian. Grateful appreciation goes out to friends and former work colleagues Andy Talbot, Mark 'Bomber' Harris, Darryl Robinson and Ian Green.

Next we need to acknowledge the generosity of those people who gave us permission to use their photographs in this book: Brendan Attard, Alistair Gardiner, Tony Hawes, Nick Martin, Baldur Sveinsson, Richard Vandervord, Darren Wilson and the one who went the extra mile for us in many ways, Mike Freer. Thanks are also due to ex-civilian AEW engineers Ron Henry and Dave Gate.

I must express my gratitude to my dad, Stan Shaw, who was the one responsible for igniting my interest in aviation all those years ago, and finally to my wife of 36 years, Tina: thank you for letting me do this and for your help, patience and stoicism along the way. I dedicate my part in the book to you.

Ian Shaw

For more information on the subject of AEW&C and all of the known AEW aircraft serial numbers you may wish to visit the author's website at: http://aewworld.weebly.com

Sérgio Santana writes:
Completing a comprehensive work on any subject heavily relies upon finding the right people willing to put their skills into the project. The following is a list, in alphabetical order, of all those who have contributed to make our book a reality:

Bill Addison, 'Akritas', Denis Almeida, Ricardo Ariel Torres Alvarez, Dalila Andrade, Shiv Aroor, David Austin, Mikhail Barabanov, Elaine Belcher, Kinder L. Blacke, Donald E. Born, Dean W. Boys, Bob Braze, Ian Brown, Geoff Cooper, Tom Cooper, Márcia B. Costley, Rebecca Danét, Gautam Datt, Rosana Dias, Anthony 'Tony' DiGiulian, Richard Engley, Robert Feuilloy, David Gibbings, Steve Ginter, Steve Gorek, Demetries Grimes, Daniele Gruppi, 'Guardião 53', Stuart Hadaway, Ross Hamilton, David Hastings, Bill Howard, Byron Hukee, Ethan Johnson, Paul Johnson, Stephen Jones, Robert 'Bob' Kison, Sven Larsson, Jim Lawler, Curt Lawson, Humberto Leite, Brooks McKinney, Sara Mejia, George Merryman, Anthony Milounakis, Peter Mills, Roger Mott, John Moyles, Thomas Newdick, Frank Offergeld, Grigory N. Omelchenko, Stuart Palmer, Kiryakos 'Kirk' Paloulian, Spiros Papadopoulos, Ian Parkinson, Marina O. Pavlova, Hadassah Paz, Jennifer Pearson, Éverton Pedroza, Druid Petrie, Bob Pisz, Anne O. Rac, Nils Marius Rekkedal, William 'Wild Bill' Richards, Lawrence 'Larry' Rodrigues, Andreas Rupprecht, Ezadi Saeed, Don Safer, Mario Serelle, Anders Skanepag, Lauren Skrabala, David A. Sloan, George E. Stewart, Henric Svensson, James Thomas, Tommy Thomason, Massimo Tiberi, 'Tphuang', Bert Udovin, Claus Wagner, Ken Walne; Ed, Sally and Pete Wasmund, Wei-Bin Chang, 'Wei Meng', Gail L. Whalen, George White, Gunther Winkle, John Wise

A number of other contributors request to remain undisclosed.

Many thanks also to Dr Heinz Berger and everyone at Harpia Publishing who joined us in the making of this book. We are also grateful to the Brazilian Air Force, Emil Buehler Library/Naval Aviation Museum, Chilean Air Force, Gatwick Aviation Museum, Hellenic Air Force, Italian Navy, National Archives (UK), Peterson Air & Space Museum, Royal Air Force Museum, Royal Navy (notably the staff of 849 NAS), and the United States Air Force.

Last but not least, the author would like to express their gratitude to his family and close friends, for their continuous and patient support during the writing of this book.

Our thanks to all of you.

Sérgio Santana

Abbreviations

AAA	anti-aircraft artillery
AAM	air-to-air missile
AB	air base
AESA	active electronically scanned array
AEW	airborne early warning
AEW&C	airborne early warning and control
ACM	air combat manoeuvring
AdA	Armée de l'Air (French Air Force)
ADC	Air Defense Command (US)
AGL	above ground level
Air Cdre	air commodore (military commissioned officer rank, equivalent to brigadier general)
AM	air marshal (military commissioned officer rank, equivalent to lieutenant general)
APC	armoured personnel carrier
ASV	Air-to-Surface Vessel (ASV), radar
AVM	air vice-marshal (military commissioned officer rank, equivalent to major general)
BAC	British Aircraft Corporation
Brig Gen	brigadier general (military commissioned officer rank)
BVR	beyond visual range
CAP	combat air patrol
Capt	captain (military commissioned officer rank)
CAS	close air support
CIC	command information centre
C-in-C	commander-in-chief
c/n	construction number
CO	commanding officer
COMINT	communications intelligence
Col	colonel (military commissioned officer rank)
Col Gen	colonel general (military commissioned officer rank)
CONUS	Continental United States
CRT	cathode ray tube
CV	aircraft carrier
CVBG	carrier battle group
CVA	aircraft carrier (attack)
CVW	carrier air wing
Det	detachment
DoD	Department of Defense (US)
ECCM	electronic counter-countermeasures
ECM	electronic countermeasures
ELF	European Liaison Force
ELINT	electronic intelligence
ENS	ensign (US Navy commissioned officer rank, equivalent to 2nd Lt)
EO	electro-optical

ESM	electronic surveillance measures
EW	electronic warfare
FAA	Fleet Air Arm (UK)
FASS	fore and aft scanner system
FIR	flight information region
Flg Off	flying officer (military commissioned officer rank, equivalent to lieutenant or 1st lieutenant)
FLIR	forward looking infrared
FM	field marshal (military commissioned officer rank)
FOD	foreign object damage
FOB	forward operating base
FTU	Fighter Training Unit
Flt Lt	flight lieutenant (military commissioned officer rank, equivalent to captain)
FM	field marshal (top military officer rank)
GCA	ground-controlled approach
GCI	ground-controlled interception
Gen	general (military commissioned officer rank)
GP	general-purpose (bomb)
GPS	global positioning system
Gp Cpt	group captain (military commissioned officer rank, equivalent to colonel)
HQ	headquarters
HUD	head-up display
I-band	electro-magnetic radiation 8 to 10 GHz
IAF	Indian Air Force
IDF	Israel Defense Forces
IDF/AF	Israel Defense Forces/Air Force
IFF	identification friend or foe
IFR	instrument flight rules
INS	internal navigation equipment
IR	infrared
IrAF	Iraqi Air Force (official designation since 1958)
IIAF	Imperial Iranian Air Force (official designation until 1979)
IP	instructor pilot
IR	infrared
IRST	infrared search and track
JFACC	Joint Forces Air Component Commander
JTIDS	Joint Tactical Information Distribution System
KIA	killed in action
Lt	lieutenant (military commissioned officer rank)
Lt Col	lieutenant colonel (military commissioned officer rank)
LTJG	lieutenant junior grade (US Navy commissioned officer rank, equivalent to 1st Lt)
LCD	liquid crystal display
1st Lt	first lieutenant (military commissioned officer rank)
2nd Lt	second lieutenant (lowest military commissioned officer rank)
Maj	major (military commissioned officer rank)
Maj Gen	major general (military commissioned officer rank)

MBT	main battle tank
MFD	multifunction display
MHz	megahertz, millions of cycles per second
MIA	missing in action
MIGCAP	MiG combat air patrol
MIT	Massachusetts Institute of Technology
MOB	main operating base
MTI	moving target indicator
NADGE	NATO Air Defence Ground Element
NATO	North Atlantic Treaty Organisation
NFO	naval flight officer
NVAF	North Vietnamese Air Force
OCU	Operational Conversion Unit
ORT	overland radar technology
OTAN	Organisation du Traite de l'Atlantique Nord (NATO)
OTU	Operational Training Unit
Plt Off	pilot officer (military commissioned officer rank, equivalent to 2nd Lt)
PoW	prisoner of war
PPI	plan position indicator
PVO	Protivo-Vozdushnaya Oborona Strany (also PVO Strany – Anti-Air Defence of the Nation, USSR)
RAF	Royal Air Force (UK)
RCAF	Royal Canadian Air Force
RCS	radar cross section
RHAW	radar homing and warning system
RNAS	Royal Naval Air Station
RoCAF	Republic of China Air Force
RoEs	rules of engagement
RN	Royal Navy (UK)
RP	Route Pack, a designated target area inside North Vietnam
RRE	Royal Radar Establishment (UK)
RSAF	Royal Saudi Air Force
RTAFB	Royal Thai Air Force Base
RWR	radar warning receiver
SAC	Strategic Air Command
SAM	surface-to-air missile
SATCOM	satellite communications
SEAD	suppression of enemy air defence(s)
SEA	Southeast Asia
SIGINT	signals intelligence
Sqn Ldr	squadron leader (military commissioned officer rank, equivalent to major)
TACAN	tactical air navigation
THK	Türk Hava Kuvvetleri (Turkish Air Force)
TRE	Telecommunications Research Establishment (TRE), UK
UAV	unmanned aerial vehicle
UHF	ultra-high frequency

UK	United Kingdom
UN	United Nations
US	United States
U/S	unserviceable
USAF	United States Air Force
USD	US Dollars
US Navy	United States Navy
USSR	Union of Soviet Socialist Republics
VFR	visual flight rules
VHF	very high frequency
VAW	Carrier Airborne Early Warning Squadron (US Navy)
VW	Airborne Early Warning Squadron (US Navy)
Wg Cdr	wing commander (military commissioned officer rank, equivalent to lieutenant colonel)
WIA	wounded in action

FIRST STEPS, 1943–1963

In April 1940, a few months into World War II, the Luftwaffe commenced its first combat operations by Focke-Wulf Fw 200C-1 Condor long-range anti-shipping aircraft against Allied merchant convoys. Operating from Bordeaux-Merignac in occupied France, the Condors sunk no less than 90,000 tons of shipping in the first two months of operations. Furthermore, they relayed positions of allied convoys by radio to the waiting German submarines. Only one Condor was intercepted during this period – by a Hawker Sea Hurricane fighter launched from a small escort carrier. These crippling attacks were soon to be declared 'the Scourge of the Atlantic' by British Prime Minister Winston Churchill.

Although they considered construction of additional radar sites on several islands with the aim of stretching the coverage of the Royal Air Force's radar network, as German bombers began reaching ever deeper over the Atlantic, the British were forced to find a different solution. In early April 1940, the scientific and military staff of the Telecommunications Research Establishment (TRE) located at Malvern in Worcestershire, determined that the American-made Consolidated B-24 Liberator was the best platform available on which to install the Air-to-Surface Vessel (ASV) Mk II radar receiver, a transmitter device, a cathode ray tube monitor (which were both still under development), and an external Yagi-type turning antenna. The whole set was to be powered by an on-board petrol electric motor.

After the initial trials had shown that even the B-24 airframe required a lighter radar set, the Aerial and Naval Interception Committee (ANIC) and the Commander-in-Chief (C-in-C) Coastal Command directed the TRE to continue developing this concept, in August 1940. The aim was to field an aerial platform with radar capable of providing coverage outside the areas covered by ground radar stations, with a range of about 80km (50 miles). While the TRE was in charge of the design and performance evaluation of the new equipment, the Royal Radar Establishment (RRE) was assigned the aerodynamic design, manufacture and installation of the rotating antenna, with the aim of – for the first time ever – fitting the transmitter, receiver, and plan position indicator (PPI) screen apparatus into a single airframe.

While RAF Coastal Command eventually took up all the available B-24s (and used them for submarine hunting in the central Atlantic), the RRE was busy developing a new, metric radar system, with the aim of developing and testing it aboard an aircraft within a year. Using a new antenna – already under development for ground-based stations with the capability to detect low-flying aircraft in sight – the RRE engineers designed a metallic aerodynamic structure, 4.6m (15ft) long, and 23–38cm (9–15in wide), consisting of eight separate elements (designed to increase the signal recep-

A close-up of R1629's rotating Yagi aerial mounted on top of the fuselage.
(via Ron Henry)

tion). This Yagi-type antenna – which lacked height-finding gear – was then installed on a rotating spindle atop a pyramid-shaped fairing on the upper mid-fuselage of Vickers Wellington B.Mk IC bomber serial number R1629, in October 1941. The rotating aerial was coupled to a reduction gearbox 'recycled' from a downed German aircraft, and powered by a 24-Volt engine, fed by a generator installed in one of aircraft's engines. It turned at a rate of 25 revolutions per minute, and could be deactivated by a hand brake. All the other main components of the radar were accommodated inside a tubular framed rack, located in the port side of the fuselage, immediately forward of the main wing spar. The controller aboard the aircraft would continuously monitor the PPI display that was a vertically mounted cathode ray tube 23cm (9in) in diameter, oriented to an imaginary line perpendicular to the longitudinal axis of the aircraft. The screen was divided into 8km (5-mile) 'range rings', with a 360° scaled calibrated outer rim, on which north was orientated to the 12 o'clock position on the screen. This became the standard configuration on all PPIs ever since.

The complete system weighed less than 320kg (705lb), which was less than 50 per cent of the originally specified weight.

Flight-testing of the modified Wellington began in February 1942. Three months later the aircraft was test-deployed to search for German torpedo boats and vector friendly fighter-bombers into attacks on them. Although the antenna's weak insulation reduced its performance by 30 per cent, R1629 was rushed into action during the night of 19-20 May 1942, this time to search for the German heavy cruiser *Lützow*, along the coast of Norway. Unfortunately, ground clutter created by high-sided Norwegian fiords prevented Gp Cpt Jack Ruttledge and his crew from finding their target.

The death knell for the metric-radar-equipped Wellington was sounded very soon after the Norwegian sortie, by the decision of the ANIC to end the metric radar program. ANIC concluded that all radars that operated above the wavelength of 10cm were 'obsolete' and insisted that all new systems operate at 10cm wavelength. Despite the strenuous efforts of Sqn Ldr Craig to apply some improvements and bring the project back to life in September 1942, the hardware was removed from R1629 seven months later. This historic aircraft was subsequently used for crew training and was written off after a landing accident later in the war.

The internal structure under the Yagi aerial was powered by an electric motor salvaged from a downed German aircraft.
(via Ron Henry)

Chapter 1

Hunting the 'Doodlebug'

Although not an outstanding success, the idea of a flying radar was not abandoned. Starting in June 1944, Germany began attacking the UK with V-1 'flying bombs'. When the Allied offensive in France overran launch sites along the coast the Germans began deploying them in an airborne launch mode from Heinkel He-111H-22 bombers based at Gilze Rijen and then at Venlo, in the Netherlands. Flying at only 30m (100ft) above the sea surface, the Heinkels were slow and sluggish, but easily escaped detection by British radars until they climbed to about 522m (1,500ft) to launch the V-1s. Although the bombers were thus exposing themselves to detection during the climb and launch phases, they were usually able to retreat safely before RAF interceptors could react. Successful interceptions thus remained rare, and were usually undertaken by night-fighters that were already airborne along the threat axis: before long, it was clear that a more effective way of detecting the approaching Heinkels was necessary, foremost one that would guarantee their destruction before they could launch V-1s.

Calculating that the latest ASV Mk III radar installed on Wellington B.Mk XIVs operated by Coastal Command in the anti-submarine role was capable of detecting the conning tower of a surfaced submarine from a range of about 40km (25 miles), and that the Heinkels reflected much more electromagnetic radiation, the RRE suggested the deployment of one such aircraft from No. 407 (Canadian) Squadron. Nicknamed 'The Demons', this unit was based at RAF Chivenor as of December 1944, when one of its aircraft was reassigned to the Fighter Interceptor Unit, at RAF Ford. Once there, the bomber was equipped with a slightly modified ASV Mk III radar, with a 71cm (28in) diameter antenna under the nose, capable of scanning 60° through the front hemisphere, and complemented by Eureka beacon (which enabled the identification of 'friendly' aircraft electronically). Flight testing proved that the aircraft was operating overweight and thus a second cockpit control column had to be installed in case one

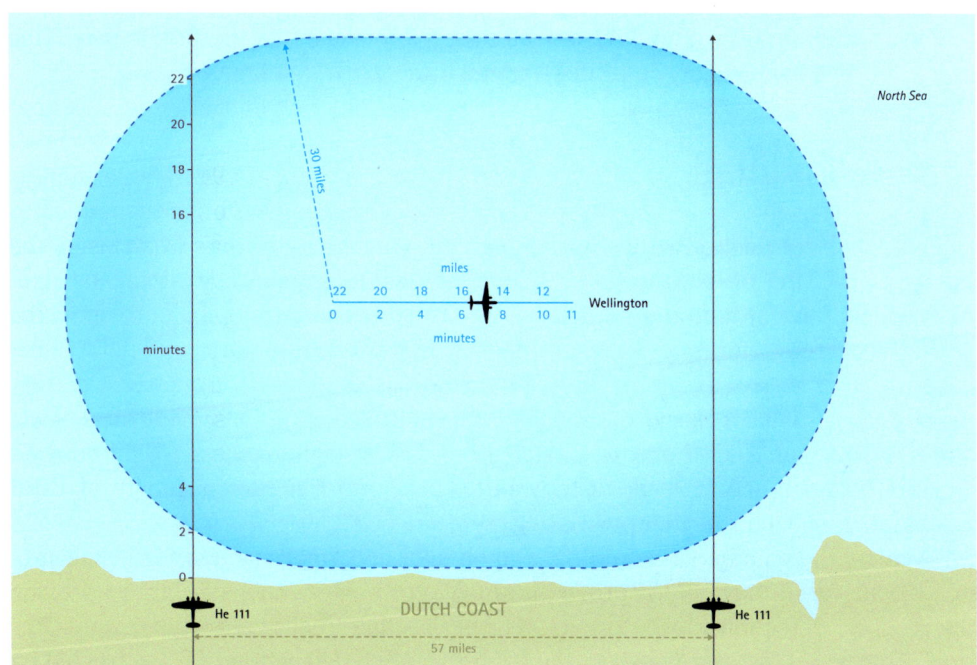

A diagram showing the flight profile of an He 111 carrying a V-1 flying bomb. (James Lawrence)

Interior of a Wellington Mk XIV showing the ASV Mk III PPI screen.
(via Ron Henry)

of the engines failed: the modified Wellington was expected to operate at only 30m (100ft) above the sea surface, and the strength of two pairs of arms would be necessary to keep it airborne if one engine quit working.

Following a period of training – Operation Vapour, which included day and night flights over the English Channel (usually in cooperation with Bristol Beaufighters that acted as both escorts and mock Heinkels) – the Wellington was rushed into service. Flg Off Ross H. Hamilton, then a wireless/air gunner serving on board this aircraft, provided the following recollection:

'The signals were sent from the antenna to a ventral dome which housed the aircraft's radar, and displayed on a screen inside the Wellington. The three wireless operators/air gunners watched the screen for one hour each, with any contact being reported to the flight crew, via intercom. The two remaining hours were spent manning the radios and the defensive machine guns. These early-warning missions lasted 10 hours or more.'

The modified Wellington flew its first operational sortie during the night of 2 January 1945, when it was sent to search for a Heinkel expected on the basis of intelligence provided by the Ultra code-breakers. It was under way at an altitude of 30m (100ft), with a de Havilland Mosquito F.Mk XXX interceptor piloted by Jeremy Howard-Williams about 800m (900 yards) in trail, at co-altitude. The two aircraft established a north-south racetrack pattern, with 11-minute legs. Maintaining a speed of 244km/h (160kt), they were able to cover an area of almost 93 square kilometres (35 square miles). However, nothing was detected, and the same remained the case over the following week. Sometime in mid-January 1945, the Canadians finally established a radar contact and vectored the escorting Beaufighter to intercept, but their target turned out to be another Wellington of Coastal Command, operating near the Dutch coast. In fact, the Luftwaffe ceased deploying Heinkels to launch V-1s on 14 January 1945, following a staggering loss of 77 aircraft and most of their crews. The Wellington experiment demonstrated that the concept of an AEW aircraft that could direct friendly fighters on to enemy aircraft was sound. It also demonstrated the lead that the British had in the area over Allied and enemy countries – a lead it was soon to relinquish.

Project Cadillac

Operating a large fleet of battleships and aircraft carriers in the vast expanses of the Pacific Ocean, the US Navy began pursuing its own ideas about development of an AEW&C platform around the same time the British began developing corresponding concepts. All relevant research was run by the Microwave Committee, officially established in July 1940 as a division of the National Defense Research Committee. Contrary to the British concept based on installation of all necessary equipment on a single aircraft, the US envisaged only the installation of radar equipment on a light bomber, whereby its radar picture would be transmitted to a control station (normally on board an aircraft carrier) for further processing.

Even so, the equipment would have to be light and small enough for installation in a navalised aircraft, and it proved a struggle to achieve this aim and to provide the required detection range. This target only became reality after the 'Tizard Mission'. This was a top-secret visit to the US by a British delegation of scientists led by

Chapter 1

Sir Henry Tizard, which took place in September 1940 with the intention of gaining access to US research and productive capacity at a time when the Americans were not involved in the war.[1]

The Americans found the British technology far ahead of their own in almost all areas, but were especially fascinated by the fully operational 'resonant cavity magnetron', a device capable of generating centimetric wavelengths essential for providing the radar with the power necessary to detect objects at long range. Originally created by the American Albert Hull in 1920 as the first form of a magnetron tube, the split-anode magnetron, the technology was developed further by both British and German scientists from the mid-1930s. The first truly effective example resulted from research carried out by British physicist John Randall, in early 1940. This included a water-cooling system (soon replaced by air-cooling), and resulted in emissions at a constant wavelength of 9.4cm. The appearance of British-designed cavity magnetrons, which were successfully demonstrated by Dr Edward Bowen to scientists at the General Electric Laboratories in New Jersey, resulted in the Americans realising that this was the key to the improvement of their own radar systems, and so rushed to put them into production with the Bell Telephone Company.

Meanwhile, the Radiation Laboratory (RL) was established at the Massachusetts Institute of Technology (MIT), and was tasked with the development of airborne aerial interception radars. Commencing in early 1941, MIT's 'Project 1' resulted in development of an operational radar that was installed aboard a Boeing B-18 bomber, in February 1941, and which proved capable of detecting other aircraft at a derisory range of 3km (1.8 miles). Eventually, Project 1 led to what became known as the SCR-720 radar, installed in the Northrop P-61 Black Widow night interceptor. Ultimately, this chain of events established the backbone of US AEW&C technology, and eventually led to the development of the radar designated AN/APS-20.

The next problem to solve was that of data transfer from the aircraft carrying the radar to control stations. In August 1942, two months after such technology would have been of exceptional importance during the Battle of Midway, the 'Indicator Group' of the RL achieved its first breakthrough, when the team led by physicist Gerald Tape developed an amplitude modulated (AM) television transmitter capable of transmitting signals from one building to the other, and then to a monitor. Nearly a year later, a highly successful test was run in the course of which an aircraft equipped with the prototype of what became the AN/APS-14 radar (which had a range of about 80km/50 miles) took place over New England, relaying its radar images to the ground station at Logan Airport, outside Boston. The system was subsequently demonstrated to a delegation of senior naval officers, but as so often in its history, the US Navy declined to accept crucially important technology on the reason that this would take up too much space aboard its carriers. The situation changed dramatically after the appearance of Japanese kamikaze suicide bombers.

As the US Navy went on the full offensive in the Pacific during the spring of 1944, its aircraft carriers found themselves exposed to attacks by the kamikaze. The high point of such Japanese operations was reached during the Battle of Okinawa, fought between March and June 1945. Although most of the Japanese raiders were either shot down or missed their targets, the few that did score hits not only caused heavy losses and material damage, but also left a big dent in the morale of US sailors. Correspondingly, the admirals now became very eager to adopt the early warning technology and the tempo of their meetings with representatives of MIT increased correspondingly, leading to a number of solutions related to the AEW&C issues. Ever better high-energy systems, identification friend or foe (IFF) systems, VHF radios, cooling systems, improved monitors and radars nearly insensitive to ground clutter began to appear, followed by a number of advanced solutions for a plethora of other problems.

Against plenty of scepticism from the admirals, the first of MIT's AEW&C systems entered pre-production testing in May 1944 within the framework of Project Cadillac, eventually resulting in the emergence of the AN/APS-20 radar, the AN/APX-13 IFF system, the AN/ART-28 Bellhop data-link, the AN/ARW-35 receiver and the AN/ARC-18 VHF voice transmitter system. This equipment was then installed in a modified variant of the Grumman TBM-3 Avenger. The torpedo bomber had proven reliability and an excellent combat record, and was roomy enough to carry the bulky equipment required. The XTBM-3W prototype made its first flight on 5 August 1944.

By March 1945, a series of 39 TBM-3Ws and related control equipment for aircraft carriers were delivered to the US Navy. Operational trials were run aboard the carrier USS *Ranger* (CV 4) between January and April of the same year, revealing that these aircraft were capable of picking up aerial targets at twice the range of radar picket ships, and surface targets up to six times further away than was possible with surface radars. However, none of the aircraft reached the front line before the Japanese surrender on 2 September 1945. Still, the US Navy had learned its lesson and continued pursuing the further development of this aircraft type. Correspondingly, the TBM-3Ws entered service with Carrier Airborne Early Warning Squadrons 1 and 2 (VAW-1 and VAW-2), officially established on 6 July 1948 and operating a total of 25 aircraft each.

The emergence of the TBM-3W had its impacts on developments in anti-submarine warfare too, because it was found that the AN/APS-20 radar could detect submarine periscopes and snorkels. Correspondingly, further sub-variants of the TBM-3W were developed during the late 1940s, which were then deployed in teams: the 'warner' aircraft equipped with radar systems were paired with 'killer' aircraft that lacked electronic equipment but were armed. Three years after their introduction to service, TBM-3Ws flew their first operational sorties during the Korean War, in November 1951, when Anti-submarine Squadrons 21 and 23 (VS-21 and VS-23) operated them from the escort carrier USS *Bairoko* (CVE 115). Reinforced by two additional squadrons, VS-871 and VS-892 (the latter was re-designated VS-38 shortly after), they continued operating in the Korean theatre of operations until December 1953, without a single loss.

Bigger is better

While successful, the TBM-3W time and again proved lacking in many aspects. Radar operators experienced serious problems while tracking specific targets in high-density areas, there were difficulties in vectoring friendly fighters, and frequent transmission interruptions between the Bellhop data-link and the command information centre (CIC) aboard the carrier. Although better than none at all, the system was obviously far from perfect. With all the problems the US Navy experienced with the kamikaze during the Battle of Okinawa still fresh in their minds, the US began developing more advanced variants during preparation for the invasion of Japan, planned for late 1945 or early 1946. The corresponding work led to an expansion of the Project Cadillac,

Chapter 1

A US Navy Grumman TBM-3W Avenger at the Naval Air Training Center, Naval Air Station Patuxent River, Maryland, circa 1946. (via Robert Sulivan)

which was to result in the emergence of a large, bomber-sized aircraft, equipped with additional communication equipment, maps, and target-tracking capabilities, which would act as an airborne CIC, capable of directing fighter-bombers in ground support operations, interceptors in air defence operations, and also provide an overall radar picture of the battle-space.

The obvious first priority of this airborne command post was to provide the capability to detect low-flying aircraft, which were expected to prove most dangerous for invasion forces. In an effort to enhance the capabilities of the TBM-3W, high priority was given to the development of a 'moving target indicator' (MTI), a device that could filter rapidly moving targets (like aircraft) from the background clutter on the radar picture. Following evaluation of different available aircraft, the decision was made to adapt the Boeing B-17 Flying Fortress for this purpose primarily because of its long range, tail-wheel configuration (providing space for a nose-mounted radome), and its wide availability (or so the US Navy thought). Thus began Project Cadillac 2.

In all, the US Navy collected 22 different B-17F and B-17G versions, 20 of which were accepted and entered the workshops of the Naval Aircraft Modification Unit (NAMU) at Philadelphia Naval Air Station as PB-1Ws, in July 1945. The main modifications included the addition of the AP/APS-20 radar in a belly-mounted radome, three 30.5cm (12in) radar scopes, large plotting boards, extra radios and new avionics that stabilised the on-board radar picture and compensated for aircraft movement. Uprated equipment included VHF, HF and LF sets, a new IFF system and radio direction finding (RDF) gear, a long-range navigation receiver (LORAN) and an instrument landing system (ILS). Due to the lack of available AN/APS-20 radar sets and the imminent ending of the war the pace of conversions subsequently came to a near standstill: the first PB-1W, BuNo 77242, did not emerge complete until 5 February 1946.

Future PB-1W crews were trained with Patrol Squadron Bombing 101 (VPB-101) at Naval Air Station (NAS) Willow Grove, Pennsylvania, and in addition to two pilots and a navigator, included a CIC commander, three radar operators, two mechanics and two electronics-technicians. This composite unit, which was also equipped with TBM-3Ws, was re-designated the Air Test and Evaluation Squadron Four (VX-4) in May

The AN/APS-20 radar installation in a TBM-3W. The antenna was installed within a fibreglass radome, located under the forward bomb bay and between the main wheels. (US Navy)

A TBM-3W Avenger prepares to launch from the catapult on the carrier USS *Franklin D. Roosevelt* (CVA 43).
(US Navy)

1946, after the Chief of Naval Operations suggested in a letter that there were still many issues and capabilities to be evaluated and more testing of both types would be necessary. Correspondingly, VPB-101 conducted a number of exercises, tests and demonstrations, further evaluating the aircraft's role in different tasks, under different weather and simulated combat conditions. This resulted in 11 reports and a 'final conclusion' provided to the Chief of Naval Operations in autumn 1949. The PB-1W proved capable of detecting fighter-sized targets over a range of 83km (51 miles) at medium altitude, but would normally use any similar sized target under way at low altitude. A trial was undertaken to reposition the radar to the upper forward fuselage, just behind the cockpit: although this configuration offered longer range and better resistance to sea-generated clutter, it was applied to only one aircraft (BuNo 77234).

The first operational PB-1W squadron became Weather Reconnaissance Squadron One (VPW-1), activated in April 1948. Based at NAS Ream Field, California, this unit was tasked with airborne early warning, meteorological reconnaissance, search and rescue and anti-submarine missions. However, it was the NAS Barbers Point-based Airborne Early Warning Squadron One (VW-1) that was to become the first to operationally deploy the type, when this unit was ordered to move to NAS Atsugi, Japan, in support of operations in the Korean War. Lt Col Bertram A. Udovin recounted the situation as follows:

'I was a flight mechanic with VC-11 [Composite Squadron 11] in April 1952, when we installed two 310-gal ferry tanks under the wings of PB-1W BuNo 77226, coded ND-8. Following system checks we transferred to NAS Barbers Point in Hawaii, on 7 June 1952 (in a flight lasting 13.8 hours), where VW-1 was formally commissioned. In October 1952 we were temporarily assigned to the Atomic Energy Commission, and did one flight taking air samples after a nuclear bomb test from Midway, on 5 November. In January 1953, I was selected to be a part of the crew of BuNo 77231/

Chapter 1

Underside view of a VW-1 Boeing PB-1W. This variant was a 'new' aircraft from a pilot's point of view, with manoeuvrability improved as a result of its modifications. (VW-1 Association)

TE-5, which was to become one of three aircraft to deploy as VW-1's Detachment A to Japan and Korea. We reached NAS Atsugi on 14 February, via Kwajalein and Guam, and few adventures. Four days later one of our aircraft due for overhauls was replaced by PB-1W BuNo 77234/TE-7, and on the 19th we flew our familiarisation flight on that aircraft, because it had slightly different flight characteristics. It turned out its equipment was better able to determine the altitude of a target, which is why – I believe – this plane was sent to Korea too. Our detachment in Japan was operated as a part of Carrier Air Group Five (CAG-5), embarked aboard USS Valley Forge (CVA 45). We flew our first mission in support of the carrier on the 25th, remaining airborne for 15.1 hours – all the time wearing cumbersome survival suits. We arrived on station over Wonsan at sunrise, checked in with the CAG-5 and proceeded, but the day was uneventful. We flew our second mission near Wonsan on 28 February and were airborne for 14.7 hours, and two more on 6 and 9 of March, lasting 12.9 and 14.9 hours respectively. All our sorties were flown in daylight.

Airborne Early Warning Squadron One (VW-1) 'Typhoon Tracers', US Navy.

'On 10 March, our crew was reassigned BuNo 77234 for 'special operations' and we transferred to the K-3, near Phang, in South Korea. The missions flown out of K-3 were quite different than those flown over the sea from Atsugi: they were shorter and we did not wear our exposure suits (though I did wear the thermal liner under my flight suit to keep warm). On these missions we usually had one or two Marine observers on board with a two-plane fighter CAP of Marine [Grumman] F9Fs overhead. We flew every day – except for some maintenance – and we often flew two missions a day. As before, these missions were for the most part uneventful. The biggest excitement I experienced was on the ground, when – while waiting for clearance to cross the runway to our plane – a bomb dropped from one of Marine fighters taking off and came bouncing down the runway towards us. The four of us dove into a ditch and the bomb bounced past us about 20-25ft [6-8m] away and came to rest on our

side of the runway about 500ft [150m] away. We stayed in the ditch until a Marine ordnance crew came and defused the bomb and hauled it away.

'*We flew our last mission over Korea on 22 March, remaining airborne for 14.3 hours and then were given another aircraft (BuNo 77231/TE-5) for return flight to NAS Barbers Point, where we arrived on the last day of that month, 1953.*

'*The modification of BuNo 77234 came at the time when the Lockheed WV Warning Star aircraft was going into production so there was no need to modify any of the other PB-1Ws. In fact the first Lockheed WV-1 was received in VW-1 in December 1952 for flight training, before we were deployed to Korea.*'[2]

Two years later, after participating in the establishment of the air defence network codenamed Lashup, which consisted of ground-based radars and AEW aircraft deployed with the purpose of detecting a possible attack by Soviet bombers, the PB-1Ws were phased out of service. Similarly, most of TBM-3Ws were retired from US Navy service in 1954.

Retrospectively, it can be said that both aircraft, and especially the TBM-3W, were very rudimentary AEW platforms. Originally designed as a torpedo bomber and rushed into service in 1942, Avenger was getting long in the tooth when modified into an AEW platform. Its primary function was little else but to transmit its radar picture via the Bellhop data-link to the CIC. On the contrary, the Cadillac 2 system resulted in an important step forward: the crew of the PB-1W not only managed the radar, but acted as an autonomous unit that could vector friendly fighters as necessary.

Foreign service

While PB-1Ws were completely withdrawn from service, many retired US Navy TBM-3Ws were subsequently sold to other nations. The French Navy became the first foreign operator of the Avenger's AEW variant when Flottille 9F was declared operational with this type on 27 August 1953. Three years later this unit took part in Operation

A Royal Netherlands Navy TBM-3W2 (ex-BuNo 91423) in the early overall gloss Sea Blue colour scheme that it wore when operating from HNMLS *Karel Doorman* (R81).
(Albert Grandolini Collection)

Chapter 1

Far left: A TBM-3W2 of Flottille 6F. BuNo 91705 entered service with the French Navy as serial number 246 in December 1953 and was withdrawn from use in January 1962.
(Albert Grandolini Collection)

Right: A Japan Maritime Self-Defence Force TBM-3W.
(Albert Grandolini Collection)

Musketeer, the invasion of Egypt by France, Israel and the UK known as the 'Suez Crisis'. Commanded by Lieutenant de vaisseau Bros, five TBM-3Ws and five TBM-3S anti-submarine variants of 9F operated from the aircraft carrier *La Fayette* (R96). Between 29 October and 5 November 1956, they flew four sorties a day on average, usually taking station some 80.5km (50 miles) away from the carrier and then controlling naval activity in the central and eastern Mediterranean. Their most important operation saw them vectoring fighter-bombers into attacks on the Egyptian Navy destroyer *el-Nasser* and frigate *Tarik*, both of which were damaged during the early hours of 1 November.

The Royal Netherlands Navy acquired a total of 25 slightly upgraded TBM-3W2s, optimised for ASW operations, in the 1950s, and the last example is known to have been withdrawn from service in 1961. The Japan Maritime Self-Defense Force also received 10 TBM-3Ws and flew them until 1961.

Early-warning Skyraider

The choice of a replacement for the venerable TBM-3W in US Navy service was practically pre-determined. In 1941, the Douglas Aircraft Company began working on the type initially known as the XBT2D-1. This aircraft evolved into the AD-1, a powerful and rugged attack aircraft better known as the Skyraider, the prototype of which first flew on 18 March 1945. During the following years this versatile design was adapted for night attack, reconnaissance, and anti-submarine operations. No less than four different AEW variants emerged subsequently, the first of which, the XAD-1W (BuNo 09107), was originally given the prototype designation XBT2D-1W.

The new aircraft was easily recognisable through the addition of a pair of side-by-side crew positions for radar operators situated within the rear fuselage, behind and below the pilot's seat. The pilot's canopy was changed from a bubble type to a faired-in projection of the turtle back that covered and cooled most of the avionics behind the pilot. Initially the AEW version had been fitted with the AN/APR-1 radar, installed inside the lower fuselage. As development progressed, that radar was replaced by an AN/APS-20, accommodated inside a fibreglass belly radome, in an arrangement identical to that of the TBM-3W. The prototype first flew in this configuration on 5 September 1947 and was soon joined by the second example (BuNo 122226) that was used for aerodynamic testing, which resulted in the addition of two finlets on the empennage to enhance the airframe's yaw characteristics in the same fashion as on the TBM-3W. Thus came into being the AD-3W, nicknamed 'Guppy'.[3]

Following extensive testing, this variant entered service with VAW-3, in late 1948, which was tasked with all-weather early warning and deployed aboard Atlantic Fleet carriers in detachments of between three and five aircraft.[3] By early 1949, VAW-1 fol-

Composite Squadron 11 (VC-11) 'Early Elevens', US Navy.

A US Navy AD-3W Skyraider. (Albert Grandolini Collection)

lowed in turn, providing similar-sized detachments for carriers of the US Navy's Pacific Fleet. The same year saw the emergence of the AD-4W, which introduced the AN/APN-2 radar altimeter (essential for carrier approach at night), P-1 autopilot and an armoured windshield. Originally designed to operate alongside the AD-3E anti-submarine aircraft, the AD-4W entered service with VC-11 (which succeeded VAW-1 in 1949), Detachment A, which participated in the Korean War while embarked aboard the aircraft carrier USS *Boxer* (CV 21), between August and November 1950. It was soon to completely replace all other variants of AEW Skyraider.

A British perspective

Following the end of World War II, British military spending experienced a massive drawdown. The situation changed with the outbreak of the Korean War on 25 June 1950.

With the Royal Navy (RN) preparing to deploy its aircraft carriers off the coast of Korea within days of the war erupting, the admirals came under pressure to improve the protection of their valuable ships. Already concerned by still-fresh experiences of kamikaze attacks on its warships, the RN was seriously worried about possibility of similar attacks off the Korean coast. The emergence of new and very fast jet fighters resulted in an urgent requirement for improved early warning, and quickly. Eventually, it was the US government that came up with a solution. In 1949, the US initiated the Mutual Defense Aid Program (MDAP), which aimed to ensure that selected US allies could support the American military fight the 'spread of Communism' through modernisation of their defences. It was within the MDAP that the US agreed to provide AD-4W Skyraiders to the Royal Navy.

The first four AD-4Ws destined for the RN were unloaded on to the dockside of port of Glasgow in November 1951, and were re-designated as the Skyraider AEW.Mk 1 on their service entry with the Fleet Air Arm (FAA). They were the first of an eventual total of 50 such aircraft. The unit selected to operate them was 778 NAS, officially established on 1 October 1951.

Pilots underwent a 90-day conversion course at Quonset Point, Virginia, followed by 20 weeks of training aboard the carrier HMS *Eagle* (R05), while observers underwent a 21-week training course. By September 1953, the FAA had received a total of 39 Skyraiders and following a number of exercises 778 NAS was reorganised as 849 NAS which consisted of six Flights (designated A through D) each of which was to serve aboard one of six RN carriers.

Generally, operations of this unit were badly hampered by a lack of spare parts, which resulted in low serviceability rates and therefore a lack of operational airframes. 'Cannibalisation' saw some airframes used as a source of spares to keep others flying.

After the final batch of Skyraiders arrived in early 1956, operating from HMS *Eagle*, A, B, C and D Flights of 849 NAS provided not only AEW cover, but also anti-submarine patrols, airborne radio relay and weather reconnaissance missions for the RN and French Navy task forces during the Suez Crisis, sometimes undertaking quite hazardous operations for such highly valuable assets, and other times serving as carrier on-board delivery (COD) aircraft. The last operational cruise for 849 NAS Skyraiders was aboard HMS *Albion* (R07), from February to December 1960. Afterwards, the last four operational aircraft were brought to RNAS Culdrose to be decommissioned and withdrawn from service.

Very early during the work-ups by FAA Skyraiders, the British decided to deploy their AEW aircraft in a manner different to the Americans: instead of serving only as airborne radar stations that transmitted their radar pictures to the CIC aboard the carrier, they decided to put radar controllers aboard their AEW.Mk 1s and let them control intercept and strike missions directly from the aircraft, the way the RAF intended to do with the ACI Wellington. Indeed, the aircraft proved capable not only of detecting enemy aircraft, but their controllers proved capable of vectoring friendly assets to intercept them. While the Fairey Gannet AEW.Mk 3 replaced the Skyraiders after only nine years of service with the Royal Navy, this point was not lost on the US Marines Corps. The AD-4Ws deployed in Korea from October 1951 until June 1952 were all assigned to Marine Attack Squadron 121 (VMA-121) and based at K-3. Contrary to the US Navy, the Marines deployed the type in support of their own AD-4Ns, deployed for nocturnal bombing attacks. Through all of this time, the US Navy continued to insist on deploying the type solely as an airborne radar, as recalled by Scott Smith:

'The scope was in the cockpit, but all the controls were in the compartment. The radar provided only raw radar returns, with no height-finding or MTI. Except for transmitting the radar picture to the CIC aboard the carrier, during one exercise we used the AN/ART-26 to transmit to a submarine under way at periscope depth, and the controller on board the submarine then vectored our fighters to intercept simulated bombers'.[4]

An AD-4W from VC-12 traps the wire on USS *Leyte* (CV 32). (US Navy)

In an attempt to combine the search-and-destroy mission of the ASW variant with that of the AEW model, in 1950 the engineers at Douglas launched a complete redesign of the Skyraider's airframe. They widened the fuselage in the cockpit area by 23cm (9in) and lengthened it by 50cm (19in), making space for an additional crewmember seated next to the pilot in the front cockpit. Furthermore, they increased the area of the vertical stabiliser by about 50 per cent, making the use of finlets unnecessary. Thus came into being the AD-5W, the later sub-variants of which had the rear cockpit hood replaced by a hump-backed aluminium cover with two small, semi-circular, blue-tinted windows. Other new equipment included an AN/APN-22 radar altimeter and an AN/ARA-25 directional homing system, but the centrepiece of this platform remained the

Skyraider AEW.Mk. 1 WT121/CU-415 (c/n 7427, ex-BuNo 12412) in the markings of the former 849 NAS, as seen at the FAA Museum, RNAS Culdrose, in 1988.
(Mike Freer)

An AD-5W of VAW-11 operating from the carrier USS *Kearsarge* (CV 33) in the early 1960s. (D. Sheley)

venerable AN/APS-20 radar, still installed in the same belly radome as on earlier variants. The AD-5N came much too late to see action in Korea, but it remained in service until 1963.

DEW Line: the great barrier

During the early 1950s, American strategists assumed that the Soviet Union was in the process of developing a sizeable fleet of strategic bombers capable of reaching targets within the Continental United States (CONUS) by flying over the icy wastes of the Arctic. Actually uncertain if any of the bomber types in service in the USSR could indeed do so, the US military assumed the 'worst case' and reacted by establishing the North American Air Defense Command (NORAD), which began developing a chain of major radar stations spread from Alaska, via northern Canada and Greenland, to Island and Scotland.[5]

With Projects Cadillac 1 and Cadillac 2 providing clear evidence that AEW aircraft could see beyond the horizon, the US Navy requested that the Lockheed Corporation develop a corresponding variant of their successful Model 749 Constellation airliner to act as a long-range, extended endurance AEW platform. The emerging prototype was designated PO-1W and flew for the first time on 9 June 1949. It was distinguished from its civilian counterpart by the addition of two large radomes that housed antennas for the AN/APS-20 search radar under the belly, and the newly developed AN/APS-45 height-finding radar on the upper mid-fuselage. About a year later, the second prototype emerged, and both aircraft were then re-designated as WV-1s. As, such, they were deployed to support the NATO exercises Mainbrace and Mariner, in 1952, where they were able to display their full potential.

Meanwhile, the US Navy went ahead with an order for a production variant based on the larger and more powerful Super Constellation, equipped with wingtip fuel tanks. A corresponding AEW&C variant entered service as the PO-2W (later WV-2 and

WV-2 BuNo 145930 from VW-1 ready to depart NAS Sangley Point in the Philippines in 1970. (Tom Harnish)

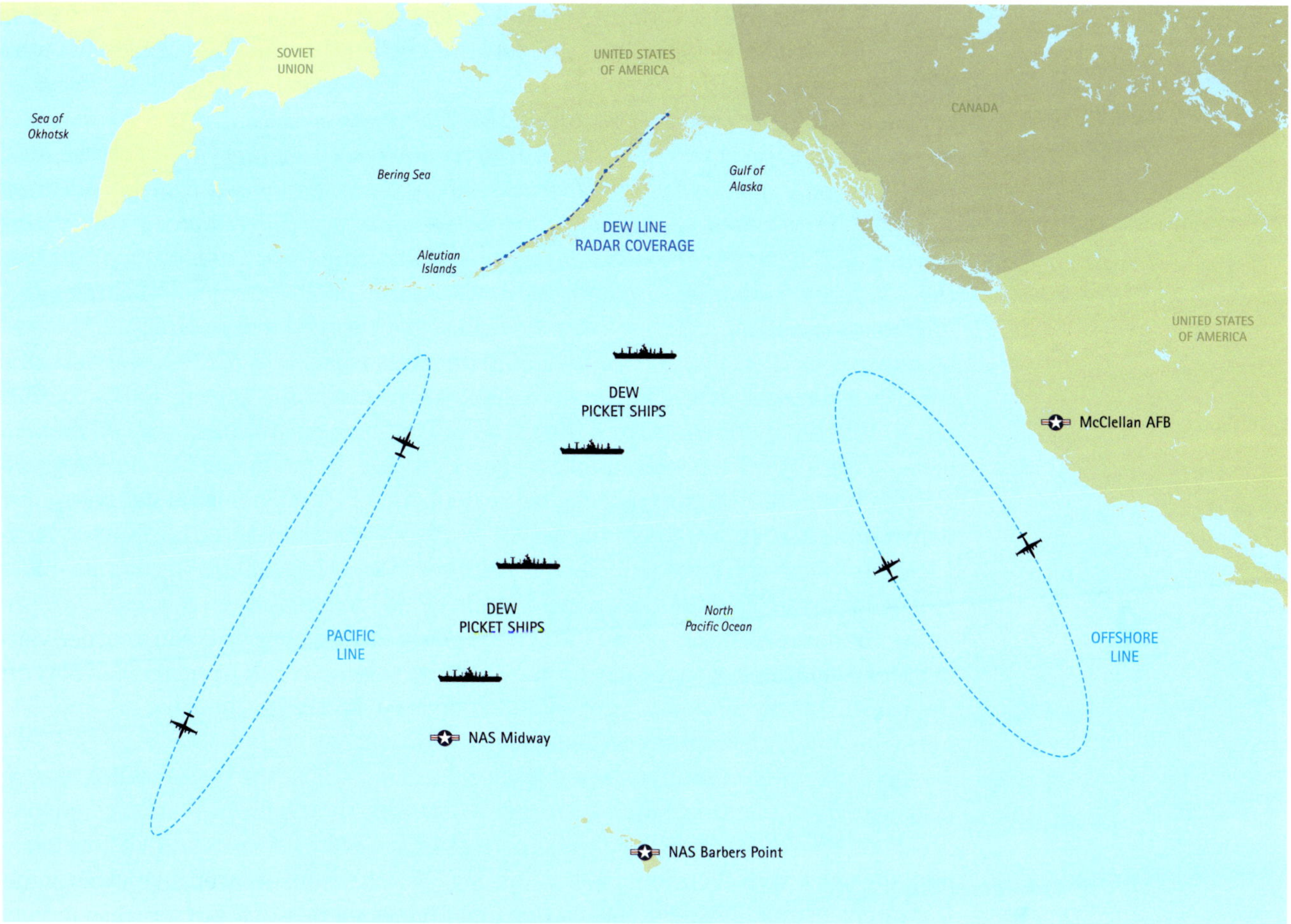

Chapter 1

A map showing the components of the Pacific Barrier. (James Lawrence)

then EC-121), and was named Warning Star. This was the first completely autonomous AEW&C aircraft, designed from the start to operate independently of support from ground stations, and to be capable of not only detecting radar contacts flying at low altitudes but also of controlling manned interceptors.

Once again, NAS Barbers Point was destined to become the major operating base of the new fleet assigned to VW-1, which was declared operational in December 1952. After several years of routine work-ups and exercises, VW-1's Warning Stars were deployed to NAS Agana, on Guam, during the Taiwan Crisis of 1958. From there, they flew missions in support of US Air Force and US Navy units deployed on Taiwan to bolster the air defence capabilities of the Chinese Nationalist Air Force (CNAF, later re-designated as the Republic of China Air Force, RoCAF), and enable it to sustain defensive operations over the garrison of Quemoy Island, which found itself under severe pressure from the military of the People's Republic of China (PRC).

The success of the US Navy's projects saw the US Air Force (USAF) eager to obtain similar capabilities and in 1951 Lockheed received an order for 10 WV-2s from the latter service. Additional orders followed soon after and no less than 73 additional Warning Stars were delivered to the USAF by 1955, many of which were actually taken over from the total US Navy order for 244 aircraft. In USAF service, the Warning Star was designated RC-121D (re-designated EC-121D in 1962). The first examples entered service with the 4701st Airborne Early Warning and Control Squadron at McClellan AFB, California, in October 1953. Jim Lawler recalled that the USAF examples were equipped with a different electronic package in comparison to US Navy examples:

'The main search radar of the EC-121D in the 1950s was the AN/APS-20, the same as on US Navy examples. But, beginning in the late 1950s and through the 1960s, the USAF replaced it with the Hazeltine AN/APS-95. The height-finding radar on the EC-121D was the AN/APS-45. Furthermore, the USAF developed the EC-121Q, equipped with the AN/APS-103 height-finder. The weather radar on all variants was the AN/APS-42, while we used the AN/APX-49 and – later on – AN/APX-101 for IFF. The most important radios sets were UHF systems designated AN/ARC-27.'[6]

Starting in 1956, and in conjunction with similarly equipped US Navy units, the USAF squadrons flying RC-121s were deployed to patrol the Distant Early Warning Line (DEW Line) off the coast of the CONUS. The DEW Line was established on both sides of the US, off the east and west coasts, as an extreme-range 'trip-wire' for protection of the CONUS from attacks by Soviet bombers. Depending on the coast over which they were flown, these 'barrier' missions, designated FAETULANT and FAETUPAC, respectively, were launched every three hours. In addition to aircraft, radar picket ships patrolled the DEW Line. Patrols along it usually lasted anything up to 14 hours (during which the aircraft would regularly cover about 5,100km (nearly 3,200 m), and usually consisted of the aircraft flying an elliptical orbit hundreds of miles off the coast, roughly between Newfoundland and the Azores on the Atlantic side, and between Midway Island and the Aleutians in the Pacific.

A typical mission profile started with the aircraft transiting to the patrol area at 2,436m (8,000ft), before climbing to 9,132m (30,000ft) for the first few orbits, and then descending to 3,480m (10,000ft) for the rest of the planned track. As well as scanning the skies with their radars, the Warning Star crews made extensive use of their IFF interrogators: indeed, these were turned on as soon as the aircraft reached the pre-planned orbit area, and their operators continuously interrogated the transponder mode C for any possible contacts for the rest of the mission. Furthermore, these military operations were supported by the civilian air traffic control (ATC), which usually relayed their data to the CIC aboard the Warning Star. In turn, the WV-2s and RC/EC-121s regularly transmitted their radar picture to respective 'barrier commanders' with the help of high-frequency radio links.

Between the DEW Line and the coast were the Air Defense and Identification Zone (ADIZ), within which interceptors of the Air Defense Command (ADC) intercepted any unidentified radar contacts – nearly always scrambling in response to calls from Warning Stars. Stations along which the AEW&C aircraft flew their orbits were randomly changed on a daily basis, in order to make it impossible for potential aggressors

EC-121T BuNo 550138 (c/n 4411) was the last EC-121D manufactured for the USAF's Air Defense Command. (Baldur Sveinsson)

Chapter 1

to predict where the AEW&C aircraft might appear at any given time. This highly intensive tempo of operations was only slowed down in 1965, when the rate was reduced to just a couple of daily missions.

Once all the air defences of the US and Canada were integrated within NORAD, they came under the control of the Semi-Automatic Ground Environment (SAGE) system, which comprised 22 command centres located across North America, and which was supervised by NORAD headquarters inside Cheyenne Mountain, Colorado, and an underground HQ in Canada. Each command centre had a pair of early 32-bit computers, capable of controlling up to 300 aerial movements simultaneously.

During the early 1960s, the USAF upgraded 42 of its Warning Stars to EC-121H standard, primarily through adding the Airborne Long-Range Input (ALRI) device, capable of forwarding data concerning unidentified aircraft to ground stations. However, the ALRI required the aircraft to climb to an altitude of about 5,200m (17,000ft) to function optimally, and this requirement put the airframes under immense stress: it is known to have caused the catastrophic loss of at least three EC-121s between 1965 and 1967. Even the CO of the 551st Airborne Early Warning and Control Wing – the only unit that operated this variant – was killed in one of these accidents.

During the second half of the 1960s, the US found itself confronted with a (much more serious) threat of attack by Soviet intercontinental ballistic missiles (ICBMs). This forced the units operating EC-121s to introduce alert and scramble techniques known as Strategic Orbit Points (STOP). Under the STOP procedure, aircraft stationed at the same base were planned to scramble within 15 seconds of each other: the first would climb on maximum power to the type's maximum ceiling, while each successive aircraft would climb to an orbit stepped down by 150m (500ft). Theoretically providing extensive coverage of the area around specific strategic locations in the US, STOP apparently did not speculate about the availability of airfields necessary for the involved aircraft to land after an ICBM attack.

The end of the airship

Following the early success of Projects Cadillac, the US Navy was eager to test its new AN/APS-20 system on as many aerial platforms as possible, to see which one might offer the best results. This resulted in AEW adaptations of the Grumman AF-2W Guardian, Sikorsky HR2S-1W helicopter, and the Lockheed P2V-5 Neptune, and then the resurrection of an older idea, which took its form in the emergence of the Goodyear ZPG-2W and ZPG-3W airships.

Considering the US Navy's history of investing heavily in the development of airships for scouting and early warning purposes in the 1930s, it is unsurprising that in 1953, the Chief of Naval Operations approached the Goodyear company with a request for the design of an airship applicable for anti-submarine warfare (ASW) purposes. After the initial design proved to be useless in this role, the US Navy decided to fit the AN/APS-20 (and the AN/APS-69 height-finder) to it, and thus came into being Goodyear's Project GZ-14.

Initially, only one prototype was ordered from Goodyear but this was later expanded to a further three airships, and another four more in 1955. The first operational airship (BuNo 4077) flew for the first time in January 1955, under the designation ZPG-2W. It

Beyond the Horizon – The History of Airborne Early Warning

A ZPG-3W AEW airship. The AN/APS-20 radar was installed within the radome underneath the gondola, and the AN/APS-69 inside the dome atop the airship.
(US Navy)

entered service with NAS Lakehurst, New Jersey-based Airship Early Warning Squadron One (ZW-1) four months later and was soon joined by five more operational examples each with a crew of 14 that were tasked with daily patrols along the Atlantic seaboard.

Before long, the US Navy began showing interest in a further-improved airship, which resulted in the emergence of the ZPG-3W. This was designed as an AEW platform from the outset. Endowed with extended endurance and improved all-weather capabilities, this model was subjected to more than 100,000 hours of static testing and an extended flight-test programme. The ZPG-3W was equipped with the AN/APS-95 long-range radar installed inside the gas envelope, and the AN/APS-69 height-finder mounted on the top of the airship, and was the largest non-rigid airship ever built. The first flight of the ZPG-3W took place in July 1958, and the type entered service with ZW-1 in September of the following year. However, only four examples were ever constructed: the first operational example, BuNo 4096, crashed during a search for a missing yacht, and the death of 18 out of 21 crewmembers together with budgetary restrictions accelerated the type's early withdrawal from service, in August 1962. The development of airships for military purposes thus came to an end for the time being.

RAF trials and tribulations

At this point we shall go back a little and look at the British approach to ground-based AEW. While the Americans were progressing well with Project Cadillac 2, the RAF had nothing comparable, not even a radar to put into a suitable platform. With the arrival of the jet age, ground-based radars on the coast could not give enough warning of jet bombers approaching. The RAF was keen to demonstrate that it could still defend British shores from the threat of the developing Soviet long-range bomber force that was starting probing flights towards British airspace.

The UK had ordered a total of 52 Lockheed P2V-5 Neptunes as a stopgap to replace its ageing Avro Lancasters in the maritime role as part of MDAP, and the first airframes arrived from the US in January 1952. The much-anticipated Avro Shackleton programme was running behind schedule and an aircraft was needed quickly by the RAF's Coastal Command to fill the gap. It did not take the British long to realise that they now had in their possession a long-range, extended endurance aircraft that was equipped with useful radar in the form of the AN/APS-20.

Fighter Command obtained three Neptune airframes, WX499, WX500 and WX501, from Coastal Command and formed them into Vanguard Flight on 1 November 1952, based at RAF Kinloss in Scotland. The aircrews and the inexperienced fighter controllers (FCs) consisted of two officers, two senior non-commissioned officers and two junior airmen that were initially trained by No. 217 Squadron and No. 236 Operational Conversion Unit at Kinloss until the-then secret Flight's airframes were ready. As quoted by Sqn Ldr C.J Lofthouse, the Flight's commanding officer, in the now declassified Secret Trials Reports:

'The intention was for the Vanguard Flight to carry out operational trials in AEW&C roles, to gain experience, and to test the effectiveness and practicability of detecting low-flying aircraft. Additionally, the trials were to serve as means of evaluation the capabilities and limitations of the equipment fitted in the Neptunes, and to prepare recommendations regarding the integration of the AEW aircraft into the communications and radar systems.'

The now declassified files on the activities of this unit show that the involved personnel tried hard to prove the Neptune as an effective AEW platform, but that the odds were stacked against them from the very start.[7]

The superior commanders' decision to withdraw Royal Navy crews, the selection of unsuitable radar operators from the ground air defence radar branch and the constant shortage of spare parts repeatedly disrupted the trials. Problems were exacerbated when the unit moved to RAF Topcliffe in May 1953, losing direct access to the testing rigs for the AN/APS-20, and being at the bottom of the food chain for the supply of spares.

Monthly reports by Sqn Ldrs Brinsden and Lofthouse tell a very clear story of how the RAF was interested in AEW, and realised its importance, but that it suffered from lack of financial resources and rivalry from other services. Examples of the problems encountered by the personnel of the No. 1453 Flight are countless, but the most important were the variable detection performance of the AN/APS-20 under operational conditions, caused by factors such as the sea state, temperature inversions,[8] aerial turbulence (which tended to severely disrupt the narrow, eight-degrees-wide radar beam, reducing the radar performance by 50 per cent and usually resulting in the loss of radar contact) and others. By July 1954 Lofthouse had become exasperated with the inconsistent performance of the radar, and desperately demanded the use of calibration sets and the secondment of their operators to Topcliffe. He had to initiate negotiations with Coastal Command for the latter to send its best technicians and best test equipment to

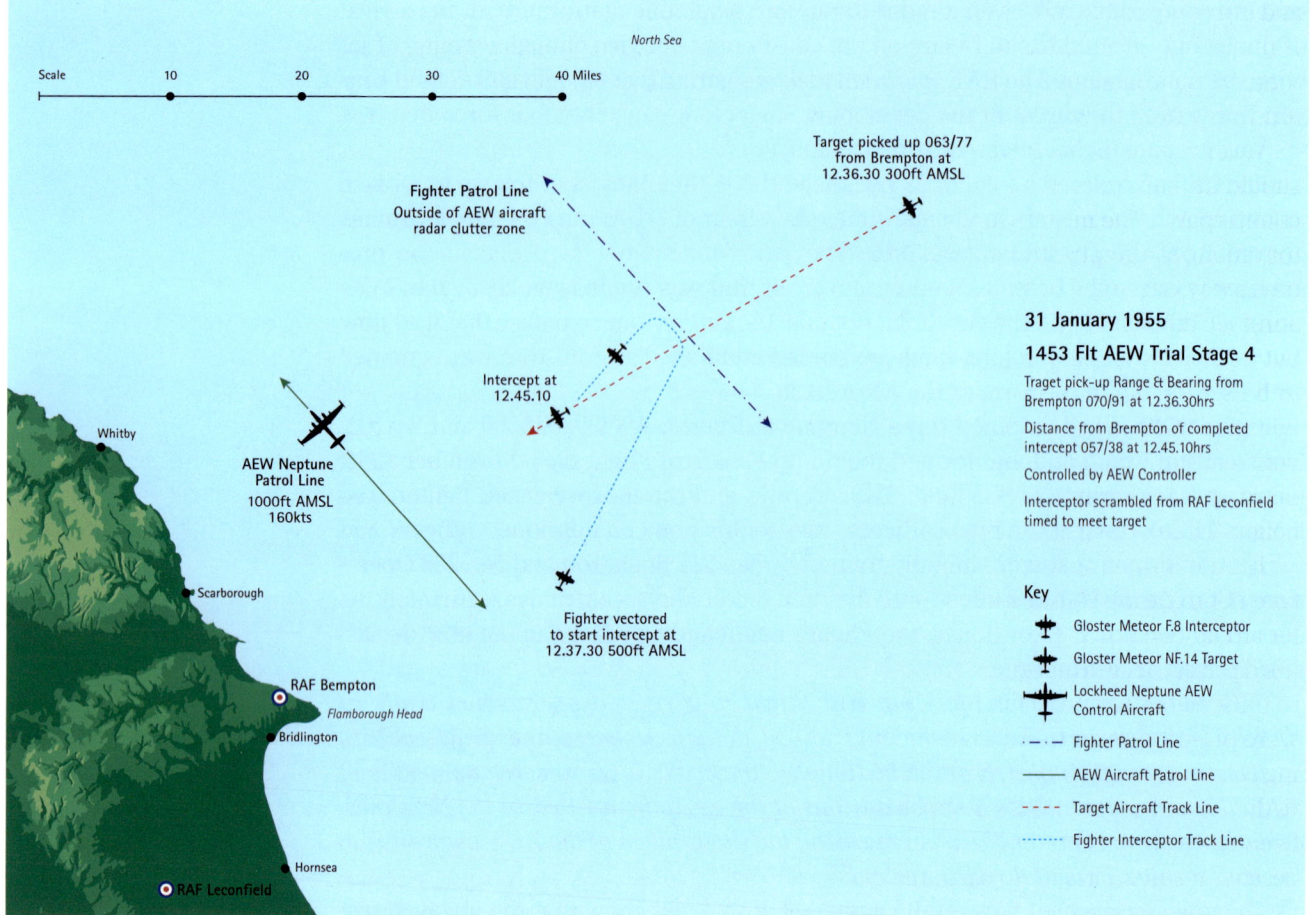

A diagram of No. 1453 Flight's AEW Radar Trial Phase 4 mission flown on 31 January 1955. (James Lawrence)

Topcliffe. It was September before No. 1453 Flight was able to report much improved results in the radar performance of its Neptunes.

In its MDAP-provided format, the Neptune was proving to be far from ideal as an AEW platform, but the money was not available to the RAF to purchase anything superior, and it had to try and utilise what it already had in service. However, as we will see later, in regards to decisions related to AEW aircraft the British government and Air Staff rarely ever got it right throughout the years. Financial constraints were the key factor, but also the commanders who insisted on forcing the trials aircraft to attempt to extend the existing coverage provided by ground-based radars rather than be allowed to operate autonomously using on-board controllers, helped to scupper the trial. Fighter Command usually ignored No. 1453 Flight's requests for fighters to act as 'playmates' altogether, forcing the Neptunes to train against each other, providing there were two operational at the same time. When fighters were available, the joint exercises were hampered by the short range of available interceptor types and complete lack of air refuelling capability. Obviously, all that this could prove was that the AEW aircraft could detect large and slow targets at an acceptable range.

Working against such odds, the crews of No. 1453 Flight continued their trials during the summer and autumn of 1954, finally getting some aid from the interceptor wing based at RAF Church Fenton, equipped with Gloster Meteor F.Mk 8s and NF.Mk 14s

Chapter 1

and Canadair Sabre F.Mk 4s. Slowly at first, they developed tactics that required interceptors to be scrambled only once a suitable target was detected. Neptunes would then transmit the airborne radar picture to the ground radar controllers at RAF Bempton who would then vector the interceptors.

Various parts of the trial dictated that the Fighter Controllers aboard the Neptunes should attempt to direct intercepts themselves without help from their ground-based counterparts. The major problem they encountered in this was the loss of the interceptor among the Neptune's own radar clutter as the fighter passed close to the Neptune on its way out to the target. The best remedy for this was found to be the fighter operating ahead of the Neptune on a patrol line well clear of the Neptune's radar clutter, but again the fuel endurance problem raised its head and patrols by the fighter proved to be too short. The controller also then had to try and turn the fighter in behind the bogey so that the fighter could engage it, and the controllers found that timing the turn was very difficult. Too soon and it did not get behind the target, too late and the target outran the fighter in a protracted tail chase. And all this was being done without any height-finding radar to establish at what altitude the bogey was approaching.

It appears that by January 1955 Fighter Command realised that it might be of advantage not to tie the Neptunes to ground-based radars, and that using the fighter controller already on board the AEW aircraft might significantly improve the effectiveness of the aircraft too. Furthermore, No. 1453 Flight finally received the necessary equipment to fine-tune its aircraft. On 24 January 1955, interceptors from the Church Fenton Wing were pre-positioned to RAF Leconfield and scrambled when an independently operating Neptune picked up a pair of No. 264 Squadron Meteor NF.Mk 14s that took off from RAF Linton-on-Ouse. Out of six sets of target runs, two were abortive (the fighters 'weren't ready to get airborne'), but the third and fourth intercept attempts proved successful, resulting in the Neptune picking up incoming targets from a range of 40-56km (25-35 miles) while flying perpendicular to the threat axis, and downloading the radar picture to the ground controllers to control the interceptions . Although the fifth and sixth runs were spoiled by deteriorating weather and temperature inversion,[8] this part of the trial actually proved that given the right cooperation and conditions the Neptunes could work as an AEW aircraft.

The trial was re-run a week later, though again with a fighter controller aboard the Neptune that was patrolling along the coast. To make the best economical use of short-ranged interceptors, a pair of Meteor F.Mk 8s was scrambled from Church Fenton to coincide with the approach times of the 'enemy' fighters. Obviously, this was

RAF Lockheed Neptune AEW Mk.1 WX547 on static display at Biggin Hill in 1954.
(Tony Hawes)

a sort of 'cheating', but it served the purpose: although the Meteors did not carry any external fuel tanks, they did run successful interceptions thanks to the Neptune. Out of eight target runs launched, only the first three were controlled by the fighter controller aboard the Neptune. One failed due to non-appearance of the interceptors. The subsequent four runs, undertaken from Bempton using the Neptune's radar picture transmitted to the ground-based fighter controller were relatively successful; a couple resulted in a tail chase after the interceptor was turned in late behind the target.

In their final summary of this trial, Sqn Ldr Lofthouse and his radar leader Flt Lt Warwick could not avoid concluding that many of the problems experienced were related to the outdated communications and relay equipment and the fact that the Neptune was not ideally suited to the role, having very small radar scopes and being very expensive to operate. Their ground crews experienced problems with a variety of spares so that on average only one out of three available aircraft were serviceable at any one time. Their radars only operated effectively over the sea, but even then the on-board controller could not see the interceptors as they passed through the Neptune's radar clutter close to the aircraft and their effective detection range was limited to only 56km (35 miles) from the AEW aircraft's position – this was, in fact, better than land-based radar coverage at that time.

Lofthouse's projection for an effective AEW force was alarming: he calculated that in order to establish and fly one patrol line off the east coast for a period of 24 hours, the RAF would require three consecutive sorties of 10 hours' duration each, which would require a minimum of two fully serviceable aircraft. To operate and sustain a series of six to eight patrol lines with each giving an extra 112km (70 miles) radar coverage over (the then outdated and soon to be replaced ground radar stations) would require a wing of 24 Neptunes operating from a single main operating base. With a total of only 50 Neptunes in service, this was obviously out of question.

Lofthouse suggested something else in his report: he proposed that in future a standard method of 'telling' from AEW aircraft to ground bases and sector controllers should comprise the range and bearing of the bogey from the AEW aircraft itself, as one would expect that the ground radar controllers would be fully aware of the friendly AEW aircraft's position at all times. As regards to equipment improvements aboard the AEW aircraft he suggested that the enhanced AN/ART-28 high-powered radar relay transmitter be fitted to give greater range and flexibility, and that the AN/ARC-28 VHF communications relay equipment be installed to enhance VHF cover, especially when directing low-flying fighters during interceptions. Foremost, he recommended the installation of the (already available) AN/APA-57A ground stabilisation equipment and the AN/APX-13A IFF interrogator to simplify the problem of control of friendly interceptors from the AEW aircraft.

Subsequently we will see how many of these suggestions were utilised in the development of the Shackleton AEW aircraft decades later.

Royal Navy success

The Skyraider AEW.Mk 1 acquired in 1952 was always seen as a stopgap measure until the British aviation industry could produce its own type. As with the Neptune, the Skyraider began experiencing severe problems with spares acquisition: an indigenous

AEW aircraft was needed, and quickly. In 1954, the Admiralty issued its Requirement AEW 154, and Fairey Aviation won the contest by offering a modified version of its successful Gannet AS.Mk 4 ASW aircraft. The resulting Gannet AEW.Mk 3 brought an extensive redesign, including a stretched fuselage, necessary to accommodate two radar operators and their radar scopes, and was powered by an up-rated Double Mamba engine, necessary because of nearly 2,500kg (5,512lb) of additional weight. The Gannet AEW.Mk 3 was still equipped with the AN/APS-20F radar installed inside in an under-fuselage radome made of fibreglass. By this time the radar was a mature and reliable system, proven capable of detecting not only airborne but also surface vessel targets and vectoring interceptors. The Royal Navy became so adept at using it to maximum efficiency that even years later RAF radar operators serving on Shackletons admitted that they never managed to get their radars to work as well as the Royal Navy's.

849 Naval Air Squadron, Royal Navy.

The first Gannet AEW.Mk 3 prototype, XL440, made its maiden flight on 20 August 1958 from RAF Northolt, and completed deck-landing trials on board the carrier HMS *Centaur* (R06) in November of the same year. The second prototype (XJ448) was the first equipped with a full avionics suite and radar equipment and it flew in late 1958. In August 1959 it was used by the Intensive Flight Test Unit, 700G NAS at Culdrose, the purpose of which was the training of crews for the future 849 NAS.

Once it entered service with 849 NAS in February 1960, the main purpose of the Gannet AEW.Mk 3 became fleet air defence and fighter direction. The squadron usually operated a four-aircraft detachment embarked aboard each one of the six fleet carriers in service with the RN at that time, while the rest of the unit was busy back at base at RNAS Culdrose with training and technical improvements.

In service, the Gannet AEW.Mk 3 proved a reliable aircraft, crewed by well-trained personnel. Although capable of carrying additional external tanks, it was usually operated without since they interfered with the radar. The Gannet could remain airborne for five hours, thus offering sufficient on-station time. The AN/APS-20F's detection range varied from 97km (60 miles) for low-flying objects, to 241km (150 miles) for objects flying at medium and high altitudes. The system still suffered from sea-surface clutter, particularly so in poor weather or when the aircraft was operating at low altitudes; temperature inversions could also wreak havoc with it, and it easily lost track of fighters with a low frontal radar-cross-section (RCS). But the operators were able to track selected targets with the help of a wax chinagraph pencil, and thus conclude whether the target was manoeuvring, or where it had reached a point suitable for interception. All contacts were shown on displays stabilised by the aircraft's Doppler navigation system that computed drift and relative speed data. The displays were coupled with a compass and as a result ground-related movement presentation was obtained, and the monitors were orientated towards the north. The system was significantly improved in the 1960s, as recalled by radar scientist Ron Henry:

'We introduced a new video accumulator ('radar amplifier'), enabling the AN/APS-20F to receive weak echo returns from ground targets, and show them on the PPI. Eventually, we developed an airborne MTI – working on the basis of the Doppler effect – and adapted an IFF interrogator for installation, and then fitted all of this into the AEW.Mk 3 serial number XL502. This helped the radar observer to differentiate between echoes from stationary ground and moving aircraft. It was a massive improvement: the radar operators could now not only track their own interceptors,

but also interrogate their and target aircraft's 'squawk' – or IFF code – in real time, and immediately knew what was a target and what not. Following testing, such improvements were installed to all the AN/APS-20Fs in service, earning them the suffix 'I', which stood for 'improved'.'

Gannets in action

FAA Gannet AEW.Mk 3s experienced the first combat deployment during operations against communist insurgents on the Malaysian Peninsula, between 1962 and 1966. Four aircraft embarked aboard HMS *Victorious* (R38) but sometimes operated from RAF Seletar near Singapore too. These Gannets primarily acted as forward air controllers (FAC) for fighter-bombers involved in counterinsurgency (COIN) operations, or control of combat air patrols (CAPs) flown by interceptors. Ian Parkinson, who flew Gannets with 849 NAS from 1964 to 1966, recalled:

'I had been slightly disappointed to be sent to 849, but I was told that my instrument-flying skills were especially good and I would be better suited to their role. Already the welcoming by squadron CO, Lt Cdr 'Butch' Barnard, made it clear that this was going to be fun. The conversion to the Gannet T.Mk 5 was easy… and I soloed after only four hours… After 50 hours of flying time on the trainer, I was introduced to the AEW.Mk 3 simulator. This was not very sophisticated (this was early 1965!), but it did have a terrain map following capability, and was all that was available to learn the differences between the two variants… Gannets were generally pleasant to fly.

We embarked aboard HMS Eagle *in September 1965, after a transfer flight from Plymouth via Brawdy to Luqa on Malta… After quick re-familiarisation and some practice deck landings at Hal Far, I did my first actual deck landings on* Eagle. *The two guys in the back must have been sweating hard, especially because my first circuit was too tight and I chose to overshoot early. The second and subsequent hookless attempts went well, though, and the first trap was on target.*

Gannet AEW.Mk 3 XL497/041 'R' parked on the flight line at RNAS Lossiemouth. To the left of the aircraft is a Palouste self-starter that could be attached to the wing pylon for transportation.
(Mike Freer)

Chapter 1

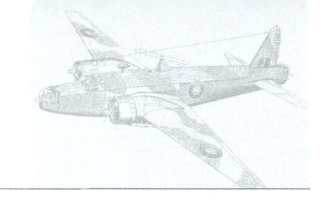

We left the Mediterranean and transited the Suez Canal, where Egyptians flew MiG-15s past us in a show of strength; not a very good show as they would often turn round in sight of the ship and return showing the other side of the aircraft with different numbers on it! From the height of the bridge we could see an airfield close to the canal with many aircraft parked, quite a lot on jacks and a few resting on their bellies on the ground.

'Once near Singapore, most of our work was related to practice exercises, controlling CAPs. I prided myself that I could always find the ship and never be more than a few miles out in my 'dead reckoning', which proved very useful because the ship and the aircraft often experienced various and simultaneous failures. On one occasion we were directing a [Blackburn] Buccaneer strike against insurgents when on the way back in formation with my CO, I found that instead of a noisy engine there was silence and I could clearly see all eight of my propeller blades. Luckily, we were still at 3,500ft [1,067m] and after some hard work I've got one engine going again… One of my observers had previously experienced a 'cold shot' [catapult failure] and narrowly escaped from an aircraft that was run down by the carrier while sinking. Understandably, he wanted to bail out rather than remain aboard for a single-engine landing, but I said it was safer to stay with us… After carrying out some tests to see how easy it would be to go around again if I made errors on landing, I made a near-normal approach at a speed of only some 70kt… We landed, picking the target wire, number 3.

In early February 1966 the ship returned for Singapore only to be re-routed in a hurry for where-we-knew-not… Nine days of steaming in a westerly direction we turned up in the Mozambique Channel and I flew a patrol off Gan, in the Indian Ocean. Afterwards we moved further south to join the Beira Patrol, the aim of which was to prevent tankers from offloading fuel for Rhodesia in the port of Beira, in support of sanctions against the illegal apartheid regime in that country. We relieved HMS Ark Royal, *which had broken down… and then commenced a 110-day patrol, flying up to five hours a day. Gannets mostly flew at dawn or at dusk, launching an hour before sunrise and landing an hour after sunset. Our role was to protect the ship and*

The Gannet AEW.Mk 3 was a remarkably manoeuvrable aircraft for its size and bulk. XL502 is seen here during a display at RNAS Culdrose in 1975.
(Mike Freer)

Beyond the Horizon – The History of Airborne Early Warning

the force from airborne intrusion in the threat direction. We would be tasked to patrol in the direction of the threat and usually about 75 to 100 miles [121 to 161km] in that direction. The radar had a range of over 250 miles [402km] (in the right conditions and height) and so we could direct CAP aircraft to intercept any incoming threat. Our heights depended on the task, the weather conditions and the range required.

'We would carry out a ship plot and direct Buccaneers and [de Havilland] Sea Vixens to fly visual identification. If necessary we would then call frigates to board the targeted ship. The area we covered was huge and the weather frequently rough with lots of thunder and plenty of cloud cover, but we experienced very few accidents.[9]

Meanwhile, Gannet AEWs deployed aboard HMS *Centaur* saw action in the course of the Radfan Campaign in the then British Crown Colony of Aden. Radfan is a mountainous area some 56km (35 miles) north of the port of Aden, with peaks raising up to 2,134m (7,000ft), hot, dry, inhospitable and populated by desert tribesmen. Local trade unions were organising strikes against the British rule and the situation deteriorated to a point where an armed insurrection erupted, in December 1963. Because of the difficult terrain, the British ran their COIN campaign primarily using air power. The RAF maintained a large and very busy Khormaksar air base, which bisected the throat of the Aden Peninsula. By 1965, no less than nine RAF squadrons flying Hawker Hunters, helicopters and various transports were based there, and RN carriers would periodically assist in combat operations, employing Sea Vixen and Scimitar fighters for ground-attack sorties. Gannet AEW.Mk 3s primarily acted as FACs, directing and correcting air strikes in close cooperation with ground forces.

Grumman's Tracer

Realising that the TBM-3W was approaching the limits of possible further development, Grumman launched the development of an entirely new AEW platform in the early 1950s. Initially given a false start and then put on a backburner, the resulting research and development process was to prove very long and drawn-out, but was eventually to culminate in the emergence of the Grumman E-2 Hawkeye, decades later.

Around 1955, a re-design of the Grumman TF1 (later C-1) Trader crew trainer/transport (itself a redesign of the Grumman S2F/S-2 Tracker ASW aircraft) emerged. This new design had the Hazeltine AN/APS-82 radar installed inside a large teardrop-shaped radome, measuring no less than 6 x 9m (20 x 30ft), installed above the fuselage. The AN/APS-82 was a significant improvement over the APS-20, although it was a much larger and heavier installation: the antenna alone weighed about 1,440kg (2,330lb) and measured 7.3 x 0.76m (24 x 2.5ft). The resulting aircraft provided enough internal space for the required electronic equipment and the crew, while still remaining small and light enough to operate from aircraft carriers. The two prototypes, originally designated XWF-1s, were ordered during the same year: following successful trials, the type was ordered by the US Navy as the WF-2 Tracer in 1957.

While the mission avionics suite underwent a detailed and lengthy evaluation process, Grumman began manufacturing a series of 40 aircraft, followed by additional batches of five, 12, and another 22 aircraft manufactured by 1959. The type entered service with VAW-1 at NAS North Island, California the same year, and two years later the re-designated and improved E-1B Tracer entered service with VAW-12. Steve Gorek,

The AN/APS-20F(I) radar scanner within its cradle and preserved underneath Newark Air Museum's Gannet AEW.Mk 3 XP226. (Ian Shaw)

Chapter 1

An E-1B from RVAW-120 'Greyhawks'. This Fleet Replacement Squadron operated the type from 1967 until 1973 from its homebase of NAS Norfolk, Virginia.
(Steve Williams)

who served as a naval flight officer (NFO) radar operator with the next unit to receive the type, VAW-11, recalls:

'Our usual working day during the carrier cruise began with extensive briefing provided by an Air Intelligence Officer, which would include all the pilots and flight officer, and cover all the necessary information for the given date and time period, such as 'hot areas', frequencies in use, coordination with other aircraft and crews etc. Following the brief, we would return to the squadron ready room to receive final pre-flight details. As the radar controller, I would review technical records of the radar, communications and control systems. With these tasks complete, the crew would head for the flight deck, perform the pre-flight check, power up and taxi for the cat. At the end of the cat stroke, the plane went into the air and then seemed to just hang there: the engines then took over from the thrust of the cat shot and off we went. That's when the work began.

'In the rear, I would first turn the radar on and warm it up. Once on station – always selected relative to the threat axis we had been given – I called the CIC to advise them of our mission readiness... Given good systems, I'd then generally check in with the aircraft assigned to me. These might be fighters or attack aircraft, some exercising air defence, others surface attacks etc. On missions with night attacks, I'd take control of fighter-bombers and run them against a picket destroyer... I would give them an initial vector to the ship and then 'walk' them in. At a given point, I'd split one off 45° then steady him up to run parallel but slightly behind the lead. I would continue to walk the lead bird in to the target and finally end with a 'mark on top' transmission. At that point he would drop flares and illuminate the ship. The trailing bird would then make the attack. It was always fun working with attack guys by night.

'After about 45 minutes or so, fighter-bombers would be detaching and checking in for recovery. We would enjoy a few quiet minutes before the next launch cycle and then repeat the exercise again. On my first cruise we usually did three such cycles... Shortly into that cruise, it was decided for safety and fuel consumption reasons that we only do a double circle. Of course, three-plus hours strapped in the seat was far better than four-plus hours! Occasionally, when things were really quiet, one could tighten all the safety straps and just go 'dead', fall asleep: I recall a few times being awakened by the jerk of the trap on the deck!'

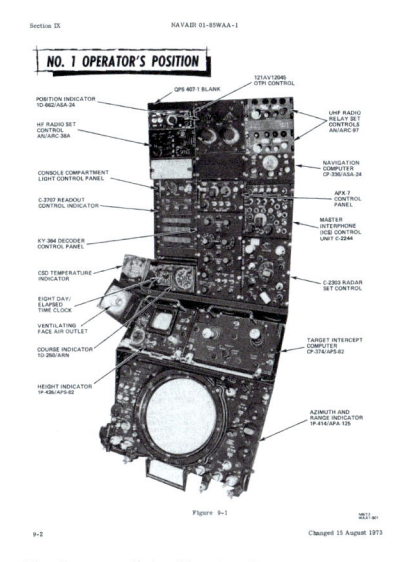

The layout of the No. 1 radar operator's position in the E-1B, as operated by Steve Gorek.
(US Naval Aviation Museum)

41

Hawkeye: vision of the future

Although proving itself as a capable workhorse in service, the E-1 Tracer was never an ideal design, but rather an adaptation of available technology to the requirement. While it was still on the drawing boards, in 1955, the United States Navy Bureau of Aeronautics (BuAer) issued a requirement for the Airborne Tactical Data System (ATDS): a system of data-links with the purpose of exchanging information between different aircraft, similar to the Naval Tactical Data System (NTDS), developed around the same time for US Navy warships.

The centrepiece of the ATDS was to become an AEW aircraft capable of loitering on surveillance station for four hours, at an altitude of 9,144m (30,000ft), slow enough for its long-range, surveillance radar to be equipped with the AMTI.

Grumman's contender design, originally designated W2F-1 and changed to E-2A in 1962, won and thus became the first ever purpose-designed AEW aircraft to enter operational service. The exterior appearance of the new aircraft, powered by two Allison T56 turboprops, has changed very little ever since. A spacious cockpit with side-by-side seating for two pilots was in front of the rear crew compartment where the combat information officer, air control officer and the radar operator were seated facing the port side of the aircraft. All three positions were equipped with 25.4cm (10in) PPIs with 12.7cm (5in) alphanumeric secondary displays underneath. Racks of avionics equipment were situated in the forward fuselage on either side between the cockpit and the crew compartment.

The heart of the complex avionics system of the E-2A was the General Electric AN/APX-96 radar that had a range of 463km (290 miles). Its Yagi-type antenna was enclosed by a large, Frisbee-shaped rotodome with 7.31m (24ft) diameter, installed on a quadruple pylon above the centre and the rear fuselage. The AN/APX-96 introduced a number of groundbreaking features, including the ability to automatically generate data on a target's course, speed and height and present these as labels on monitors, freeing operators from the need to continuously track the target and update the plot. The supporting electronics was capable of simultaneously tracking 250 targets, and supporting operations by 30 different interceptors with help of the data-link.

One of just four TE-2A crew trainers, serving with RVAW-110 'Firebirds'.
(Mark Karlosky)

The first Hawkeye prototype (BuNo 148147) made its maiden flight on 21 September 1960, but subsequent development proceeded at a relatively slow pace. The first unit declared operational on the type was VAW-11, on 19 January 1964. It was only the US involvement in the Vietnam War that accelerated further development of the Hawkeye, which eventually turned it into the AEW&C platform that it is today.

Cuban Missile Crisis and Soviet interest

US Navy units operating Tracers were among the many types put on alert during the Cuban Missile Crisis, in late summer and autumn 1962, and caused by Moscow's deployment of ballistic missiles to Cuba. Distributed into three-aircraft detachments aboard four aircraft carriers USS *Enterprise* (CVAN 65), USS *Independence* (CVA 62), USS *Lake Champlain* (CVA 39) and USS *Randolph* (CVA 19), they provided valuable assistance in support of the naval blockade Washington imposed upon Cuba. Furthermore, ADC's EC-121 Warning Stars supported operations by USAF reconnaissance aircraft over and around the island. To their surprise, it was then that USAF crews discovered that their IFF interrogators were capable of reading codes of the Soviet-made SRO-2 IFF transponders installed on MiG-17 and MiG-21 fighters operated by the Soviets on behalf of Cuban: this piece of crucially important intelligence was to prove of immense worth in air combats over North Vietnam, nearly a decade later.

The Cuban Missile Crisis left a lasting impression upon the Soviets, too. While monitoring US operations around and over the island, senior Soviet commanders became aware of how important the AEW aircraft had become for the USAF and the US Navy, how they enabled them to see further and better than any ground-based radars could, and vector their interceptors and fighter-bombers in a more effective fashion than ever before. Although the Soviet Navy did not have any aircraft carriers, it was developing a strong and effective surface fleet that could well need support from similar aircraft. The Soviet Air Force was in the process of moving in a similar direction since 1958, when it launched two primary projects related to the National Air Defence Modernisation Program: one for a long-range interceptor, which eventually resulted in emergence of the Tupolev Tu-128; and one for an AEW platform.

The Soviet Union had similar air defence requirements to the US and Canada; however, the threat it was facing was far more serious than that of its opponents, because supported by a large fleet of Boeing KC-135 Stratotankers the huge fleet of Boeing B-47 Stratojet and B-52 Stratofortress bombers of Strategic Air Command was indeed capable of reaching all major cities in the USSR. Moreover, the US bombers could do so by crossing the Arctic and then by penetrating sparsely populated northern Siberia, where they could circumnavigate the sparse ground-based radar network. To make matters worse, the USSR was already lagging behind its opponents in development of the necessary advanced radar technology: while the US had been deploying AEW platforms operationally for nearly 10 years by the late 1950s, the Soviets neither possessed the necessary technology, nor experience in operating such aircraft.

Corresponding operational requirements for an AEW platform demanded by the Soviet Air Force called for an aircraft capable of remaining airborne for between 10 and 12 hours, reaching an altitude of 12,000m (36,500ft) and patrolling an area 2,000km (1,242 miles) away from its base. It was to carry an improved variant of the P-30 Kristall

Carrier Airborne Early Warning Squadron 11 (VAW-11) 'Early Elevens', US Navy.

A Tu-126 takes off from its home base of Siauliai in the Lithuanian SSR. (Archive Beriev OKB via Yefim Gordon)

air defence radar (ASCC codename Big Mesh) named Liana, and be capable of transmitting its radar picture to ground stations equipped with the Livna communication suite. The Liana AEW radar had a 10m (30ft) long and 1.8m (5.4ft) wide antenna that generated a four-millisecond pulse of 2MW power, emitted at a frequency of 500Hz. It enabled detection of moving targets against heavy sea clutter, and presented a filtered picture with the help of wobbling the pulse frequency of the returning echoes. The system could reportedly detect bomber-sized targets out to a range of 300km (186 miles), and simultaneously track up to 14 such aircraft. In order to provide support for naval forces, it was further upgraded with ability to detect surface vessels like US Navy aircraft carriers out to a range of 400km (249 miles).

After some testing of the Tu-95 bomber as a suitable host airframe, the choice fell on the Tu-114 airliner, which had the same wings, engines, undercarriage, and tail control surfaces, but offered a spacious and pressurised cabin with enough room for avionics and the crew. Originally designated Product L by the Tupolev construction bureau, the resulting aircraft was equipped with a large rotodome installed on a single pylon mounted above the rear fuselage. The former passenger cabin was occupied by five compartments, including one for radar operators, aerial intercept controllers and countermeasures operators; one with space for crew (including a toilet, wardroom and several bunks and tables); a section with the Liana's rotating gear; a communications room (including equipment for the Arfa data-link system; and the rearmost compartment for a tail-gunner that was never used operationally. Series production was ordered in May 1960, and the prototype Tu-126 (c/n 618601) made its first flight on 23 February 1962 from Bezymyanka air base.

During the State trials and test phase, which lasted until December 1964, it was discovered that the aircraft's centre of gravity had moved to the rear because of the rotodome's weight of 12,000kg (26,455lb) and aerodynamic drag, and that the latter reduced the maximum speed to 530km/h (286kt) at low altitude. Trials also highlighted problems related to electromagnetic incompatibility between more than 20 different emitters installed around the airframe. Nevertheless, the aircraft proved compatible with ground and sea command, control and communication bases, and was ordered into production for the newly established Air Defence Command (PVO). The first

Chapter 1

A Tu-126 is shadowed by a US Navy A-4E from VA-45 'Black Birds', based on the carrier USS *Intrepid* (CVS 11) in the early 1970s.
(San Diego Aerospace Museum Archive)

Tu-126 officially entered service with the 67th Independent Airborne Early Warning Squadron on 30 April 1965, at Montchegorsky, but subsequently moved to Siauliai, in the then Soviet Republic of Lithuania. From there, this unit operated in two four-ship flights, primarily tasked with early detection of US and British bombers as far as 1,000km (621 miles) outside Soviet airspace.

Although often underestimated by Western observers, the Tu-126 was, for its time, a very sophisticated and highly effective aircraft. Standard operational procedure required the radar to be switched on immediately after take-off, to get properly warmed up before the aircraft levelled of at its normal operating altitude of 8,000m (26,000ft). Data from the Liana radar was presented on video maps that used the polar coordinates system (and displayed both the azimuth and range), through the target's position was calculated using the Cartesian coordinates system. The radar could be alternated between passive and active modes in order to avoid detection of its activity by the enemy: instead, the crew could just 'listen' to enemy electromagnetic emissions. The aircraft was designed to orbit in the centre of an imaginary box measuring 1,200 by 1,200km (746 by 746 miles), with its radar sweeping the area to a distance of 680km (423 miles). The operators were able to detect the nature of radar contract (airborne or surface vessel), determine their IFF and the quantity. Furthermore, because the Livna was a bi-dimensional radar, it necessitated no separate height-finder: the target's altitude was obtained through calculators installed in the radar consoles, using the time-measurement taken by the echo to return to the antenna and taking into account the aircraft's altitude. All of this data, and up to four targets, could be transmitted in the form of a telegram containing up to 32 digits to the ground command post with help of the Vozdukh-1 data-link. A single Tu-126 was capable of simultaneously controlling up to 10 Tu-128 interceptors against as many targets with the help of the Arfa system's 1-RSB-70/R-807 radios, which relayed directional commands to the Tu-128's R-832M radio, which in turn presented the data to the crew through the Put-4P navigational system. Correspondingly, although the crews of the 67th Independent Airborne Early Warning Squadron did train with other interceptors, the Tu-128s remained their favourite partners for the duration of the Tu-126's service. Usual operational areas were the Baltic, Barents and Kara Seas, where 8.5-hour sorties were flown, during which the crews slip-changed every three hours (longer periods of rest were useless due to noise and resonance created by the turboprop engines, avionics and their uncomfortable VMSK waterproof flight suits).

Although participating in one major exercise in the Indian Ocean, in April 1970, for most of its career the Tu-126 primarily operated along the Norwegian and Swedish coasts, where its radar performance was badly affected by ground clutter: like contemporary Western AEW platforms, it lacked a true 'look-down' capability and could not detect any aircraft flying at low altitude. This was probably one of the crucial reasons for the type's retirement in the mid-1980s.

1. Ironically, the Tizard Mission resulted in a giant leap ahead for US military technology, enabling it not only to launch manufacture of war-winning equipment, but also equipment that was to establish the US as leading arms manufacturer ever since. On the contrary, the British not only paid dearly in the loss of post-war commercial profits, but also spent the next 60 years repaying debts from Lend-Lease agreements to the US government.
2. Udovin, interview, via email, 17 February 2012. Prior to joining PB-1W operations, Udovin served as the squadron air intelligence officer with VP-22, when that unit operated P2V Neptunes from NAS Barbers Point. Among others, he served on 'top secret' missions over the Formosa Strait out of NAS Naha, Okinawa (Japan) and NAS Sangley Point (Philippines).
3. Among many interesting 'postscripts' to the testing of the 'Guppy' – a nickname based on the aircraft's similarity in appearance to a pregnant guppy fish – is the one experienced by Lt Cdr Wheems during a test flight out of Patuxent River on 15 October 1948. Wheems ran into difficulties when his fuel tank turned out to be empty. Undaunted, the pilot attempted to ditch the aircraft on the surface of the river, only to find out that the belly radome made the aircraft skip across the water surface almost like a seaplane, and allow him to safely slide on to the river bank. The subsequent Board of Inquiry discovered that the radome shape had caused a reduction in the air pressure that had sucked fuel out of the fuel vent pipe, resulting in fuel starvation. Needless to say, rectifying modifications were rapidly introduced to subsequent AD-3Ws while these were still on the production line.
4. Smith, interview, 7 April 2008. Smith flew AD-4Bs, AD-4Ns, AD-4Qs and AF-2Ws and served as an instructor for the AN/APS-20 radar. Later he served as CO VA-25, when the unit converted to the Vought A-7 Corsair II.
5. Based on data that became available following the end of the Cold War, the latest studies have shown that not one of the early strategic bombers put into service by the Soviet Union was capable of reaching CONUS and returning to base – even with support from tanker aircraft. Some types, like the Tupolev Tu-95, were capable of doing so only on the condition of expending all the fuel they could carry, which meant that if sent on such a mission, and provided they were not shot down while approaching their targets, they would run out of fuel shortly after, and still well over CONUS.
6. Lawler, interview, 15–17 May 2008.
7. Lawler served on USAF EC-121s as a communication master sergeant.
8. RAF 1453 F/S. 16/3/Air Feb 55 and MOD Fm 540 No 1453 Flt
9. A temperature inversion occurs where warm air sits above cold air, or vice versa. When radar beams hit an inversion layer it can cause a form of radar clutter known as 'anaprop' that obscures the primary radar contact.
10. Parkinson, interview, via email, 3 March 2012. Regrettably, shortly after returning to the UK, Parkinson injured his back and was medically downgraded. He continued to fly helicopters, but never again experienced fixed-wing carrier operations. Notably, C Flight of the 849 NAS was awarded the prestigious Boyd Trophy for its outstanding performance during participation in the Beira Patrol from HMS *Ark Royal*, in 1966.

Chapter 2

GOING TO WAR, 1964–1968

Cold War by proxy: Vietnam

The first major conflict to see widespread and intensive involvement of AEW platforms was the Vietnam War. Following the French withdrawal from Indochina and the independence of Cambodia, Laos, North and South Vietnam, the latter country found itself facing an insurgency from the Viet Cong (VC) guerrillas, supported by North Vietnam, whose aim was a reunified country.

The Pentagon established the Military Assistance Advisory Group (MAAG) in South Vietnam, in 1956, and began providing training and arms to the South Vietnamese armed forces, but they proved ineffective. By January 1964, US Navy warships began patrolling the North Vietnamese coast in an attempt to track down some of the vessels that were transporting supplies for the VC, and related air reconnaissance operations were supported by E-1B Tracers of VAW-11's, Det C, embarked aboard USS *Kitty Hawk* (CVA 63).

Tracers were tasked with performing the entire spectre of their repertoire, from fighter control, search and control of surface contacts, and support of search and rescue (SAR) operations. The mountainous terrain of South Vietnam made the identification of targets flying over land very difficult and required a high degree of skill from their radar controllers, but generally, the type proved robust and reliable. Steve Gorek recalled a mission flown from USS *Coral Sea* (CVA 43):

'The E-1B was a stable and extremely reliable AEW platform and a bird that got the job done with little fanfare or sophisticated systems. One evening well after dark, on a very black night, I was airborne in an orbit covering the usual launch cycle. My radar systems were good and things were pretty quiet. I was in communication with both the ship and the aircraft under my control but what was very unusual was the ship's communication. Normally, the ship was crisp, clear and strong but that night it was heavy with static, intermittent with consistent cutting in and out. Communications continued between the ship and the aircraft launched that cycle although one could tell things were not normal with the number of 'say agains' being requested.

'Occasionally I would be in contact with the ship but finally during a string of transmissions, the ship simply went dead. There was no transmission activity at all, but of a greater urgency there was no navigational signal; the TACAN went dead as well. The situation was such that all aircraft were flying around with no reference to the ship, no ability to contact the ship and, of course, burning fuel by the second. After numerous attempts to contact the ship, I broadcasted an emergency *call signal and had all aircraft check in with me, as we were the AEW bird. Each individual aircraft was asked to report their fuel state.*

'Obviously, we had the advantage of our radar and were able to 'paint' the coast of the Philippines, plus our multiple radios as an airborne control centre. Everything worked fine and once all the aircraft had checked in, I confirmed that the tanker was in orbit overhead the ship. Confirming this, I directed all aircraft to home in on his signal and directed him to transmit a long count on the frequency so the other pilots could home onto him. I also began assigning altitudes to the various sections, but on the pilots' suggestion, simply had them ascend/decent to their assigned emergency marshalling attitudes. The aircraft complied with attitude adjustments while homing on to the tanker. As the aircraft began to close overhead the USS Coral Sea, *I provided 'bingo' vectors to NAS Cubi Point [in the Philippines]. Fortunately, just about that time* Coral Sea *comms came back online. What a blessing! As it was getting close to the end of the cycle, the situation for the jets was getting critical; recover quickly or divert with 'bingo' fuel to NAS Cubi: fuel would be tight.*

'In the end, all aircraft recovered safely to the carrier and as usual we were the last to come aboard. On getting down to the ready room, I was immediately surprised and embarrassed, as it seemed I was the hero of the moment for stepping in and taking control of the cycle when carrier comms went down. Despite the embarrassment, it was a very satisfying moment, although a very busy evening.'[1]

The apparently calm situation off the coast of Vietnam experienced a dramatic change following the 'Tonkin Gulf Incident', on 2 and 4 August 1964: after two US Navy destroyers supposedly came under attack from North Vietnamese torpedo boats, Congress ordered retaliatory air strikes to be launched against North Vietnam the very next day. The US thus officially entered a war its forces had been clandestinely fighting since February 1960.

The pace of air warfare increased throughout 1965, when the USAF began deploying tactical fighters to bases inside South Vietnam and Thailand. Washington then ordered them to strike additional targets inside the North Vietnamese 'panhandle', while the US Navy established a near-continuous presence of three carrier battle groups (CVBGs) in the Gulf of Tonkin, on Yankee Station (north) and further south on Dixie Station.

Because the US government was worried that China would become directly involved if the enemy was bought to its knees, all targets were selected and vetted by the White House, which severely restricted the effectiveness of air strikes on the North and hamstrung the US military planners. This was also done in the vain hope of keeping the North interested in continuous peace talks. The administration of President Lyndon B. Johnson thus imposed strict sets of rules of engagement (RoEs) and Defense Secretary Robert McNamara maintained strict oversight on every single operation.

When the US Navy began launching its regular strikes against selected targets in the North, in March 1965, there was virtually no coordination between its and similar operations undertaken by the USAF. Because of this set of circumstances, the main task of US Navy AEW aircraft remained the security of Task Force 77 (TF-77) and its warships under way in the Gulf of Tonkin.

Similarly, when the USAF began deploying its EC-121D AEW aircraft over Vietnam, they did not link up with any of the US Navy assets. Furthermore, because early US naval aviation operations against the North took place during daylight hours only, the AEW aircraft seldom operated by night, leaving the ships of TF-77 vulnerable to possible nocturnal attacks. A solution was found through the US Navy deploying its VW-2 Warning Stars from NAS Agana, Guam, to NAS Sangley Point in the Philippines, so that they could fly AEW missions over Tonkin by night. Later on, Det C of this squadron deployed to Da Nang AB, in the northern part of South Vietnam, from where it operated until its withdrawal in 1967. Pete Wasmund served as an electronics crew chief with VW-1 around that time:

'I am actually a naturalised American citizen: my parents were British. I was naturalised on the island of Guam at the request of the CO VW-1, to ascertain the necessary security clearance. Our unit was not a 'barrier squadron', but was tasked with tracking typhoons until we were ordered to provide support for TF-77. I only spent 16 months with the squadron but these became the most memorably in my four-year career with the Navy. Our mission over the Gulf of Tonkin was to perform air and sea surveillance by night. First we flew at around 1,500 to 3,500ft [457-1,067m], just north of Da Nang, but our patrol areas then began migrating further north and east. By the end of 1965 or early 1966, most of our missions were north of the 19th Parallel and many even above the 21st Parallel, between the coast of North Vietnam and the Chinese island of Hainan.

'One particular mission I recall very well was flown in early 1966. We picked up a bogey at 320 to 330° on the compass. It was well inland over North Vietnam but heading in our direction and about 110-140 miles [177-225km] away. We reported it to 'Alpha Whiskey' [the US Navy's central air traffic control for airspace over and around North Vietnam, also known as 'Red Crown'], *but he was not concerned. Still, we kept tracking that contact and one of guys handling the ECM equipment identified it with no doubt as a MiG-17, perhaps even a MiG-21. We continued tracking the MiG and realised it was on a direct course for USS* Enterprise. *He crossed the coast and reached the Gulf when Alpha Whiskey began questioning if we were sure we had a contact. We confirmed that it was and our conversation continued for a few more minutes. Finally, we had that MiG within 60 miles [97km] of the 'Big E', when they finally launched a pair of* [McDonnell Douglas] *F-4 Phantoms. Within 10 seconds, the MiG did a 180° turn and headed back to North Vietnam. To this day I keep wondering if this was a mistake or if someone was trying to sucker that MiG close enough to be able to shoot him down...'*

Pete Wasmund served as an electronics crew chief with VW-1 in 1960–1962 and 1963-1966 and recalls the equipment inside the US Navy's Warning Stars and their early operations over Vietnam:

'Each flight crew performed most of the maintenance on its aircraft, which gave us a very comforting feeling, knowing that the man who maintained the equipment were also on the flight. My primary responsibilities were repair of electronic equipment on the ground and in flight. If we could not get the equipment repaired in flight we would haul it to the shop, repair and re-install it and then check it out in the aircraft.

'Primary search radar of our EC-121s was the AN/APS-20B/E. It had a maximum range of about 250nm [463km], but using the 'delayed sweep' mode, it could reach out to almost 300nm [556km]. Peak power output was about 2.5MW. The antenna could be selected to rotate at six turns per minute, two turns per minute, or for sector scan. It was tilt-stabilised. The APS-20 could be selected to transmit in high pulse repetition frequency (PRF): this limited the max range to 50nm [93km], but pro-

Airborne Early Warning Squadron Two (VW-2), US Navy.

Radar controller consoles and CIC on board an EC-121. (VW-1 Association)

vided better resolution. With the help of the AN/ART-28 data-link we could transmit our radar picture to ground stations, aircraft carriers and to other aircraft.

'We carried many spare parts for the APS-20 radar system in the aircraft and could carry out extensive repairs in flight. The spare parts kit included all of the various vacuum tubes for each piece of electronic equipment on board including a spare magnetron and hydrogen thyratron for the APS-20. On some missions we worked several hours coaxing the radar into operation. On one occasion we used paper coffee cups scavenged from the trash bins and with their bottoms cut out and taped them together to make a long tube to supply cool air to the radar cooling unit because its blower motor failed.

'The AN/APS-45 radar system as used in the EC-121 was the height-finding radar. It had a maximum selectable range of 120nm [222km], and radiated a pattern that was 1.5° wide in vertical and 6° wide in horizontal axis. When the CIC operators requested the altitude of a particular target, the APS-45 would receive target range and bearing information from the CIC operators' cursor position, which repositioned the radar antenna to the bearing of the target and placed a range strobe at the target range. The APS-45 operator could make minor adjustments to the antenna bearing and place a height cursor over the target and send the information back to the requesting CIC operator. A modification installed in the mid-1960s allowed the CIC operators to see the height-finder radar PPI picture on their displays. However, this was of very limited usefulness, probably only used for weather and not target tracking. The system was there but I don't recall it being used on any missions I was

Chapter 2

on. The majority of the maintenance problems with the APS-45 were related to its antenna, which was 99 per cent dependent on functioning vacuum tubes.

'Work of both radars was displayed to operators by the AN/APA-56 display system, which was also 99 per cent dependent on vacuum tubes. Each aircraft initially had five consoles installed, given the designations ACO 1 through 5. Each ACO received information from the AN/APS-20 system, the AN/APX-7 IFF interrogator system with SIF (selective identification feature) decoder, the AN/APA-57 ground stabilisation equipment and the AN/APS-45 height-finder. Each operator had access to five separate UHF radios and could communicate via the Internal Communications System (ICS) to any other position in the aircraft. A camera scope was also a part of the AN/APA-56 system. A 35mm camera was attached by a special mounting to be focused in on a relatively small CRT. Adjacent to the CRT was a small clock and a notepad that were included in field of view of the camera. These were used to record the time and other information that would be pertinent to the pictures being taken. The camera could snap a picture at a selected number of antenna rotations. As I recall this was used most often during our weather missions.'

'The AN/ASA-14 Dead Reckoning Table (DRT) as used in the EC-121 was used by the CIC officers to give an overview of all targets or weather patterns in the area. This system received information from the Ground Stabilization Equipment (AN/APA-57), which drove an X-Y plotter with a pinpoint light beam, which was directed to shine through the plastic tabletop. This beam showed the position of the aircraft on a map, which was laid on the plastic tabletop of the DRT. All targets would be reported to the DRT operator and plotted with all available information on the map. This table was also used extensively during weather missions. The 'Green Worm' was the common name given to a radar altimeter that displayed a circular sweep pattern ('J' scan) that showed the transmitted pulse and received echo. A scale on the outer edge of the CRT would tell the exact altitude of the aircraft by comparing the barometric altitude readings and the radar altitude readings the surface missions.

'From 1963 to 1966 the squadron only had seven EC-121K aircraft and one C-121J with a primary mission of weather reconnaissance and a secondary mission of AEW for the 7th Fleet. VW-1 handled the night-time storm fixes at 09.00 and 15.00 and the US Air Force's two weather squadrons took the daytime coverage, 03.00 and 21.00. VW-1's C-121J was kept busy flying parts and people all over the Pacific.

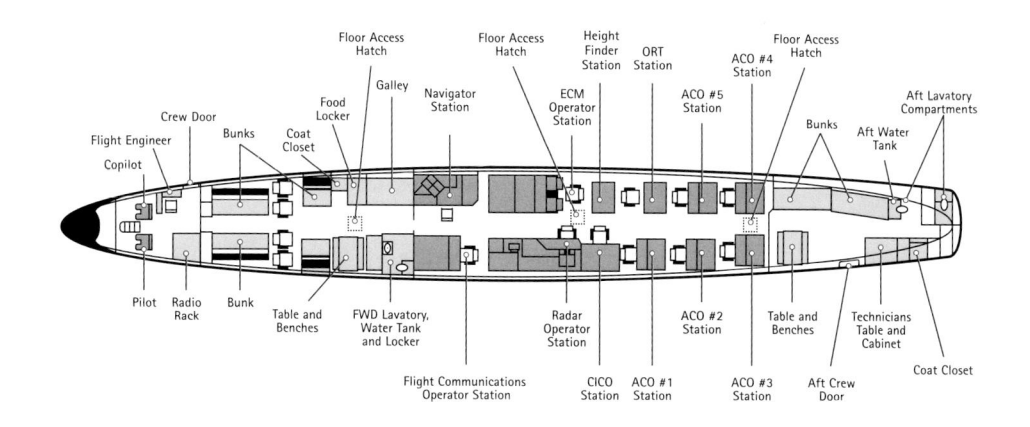

The layout of crew positions in the WV-2 Warning Star. (James Lawrence)

'After the Gulf of Tonkin Incident, VW-1 entered a period of much heavier workload. We were tasked with providing night-time (23.00 to 06.00) AEW coverage to 7th Fleet units operating off the coast of North Vietnam. To provide this coverage VW-1 established a two aircraft detachment that operated out of either NAS Sangley Point or NAS Cubi Point. The total flight time of each mission was typically 15+ hours. Each aircraft and crew would fly alternate nights providing AEW coverage for the fleet, then return to Agana after 14 days with the detachment, to perform our missions of typhoon reconnaissance.

About a year after I left the squadron (in August 1966) the detachment in the Philippines was moved to Da Nang and then Chu Lai [both in South Vietnam]. This reduced the transit time to and from the operating area from nearly eight hours to about two hours. VW-1 then started providing 12-hour AEW coverage to the fleet.'

Delayed Hawkeye

The escalation of hostilities in Southeast Asia and the need to replace the Tracer forced Grumman's hand and the E-2 Hawkeye was rushed into an inaugural cruise aboard USS *Kitty Hawk* from October 1964 until June 1965. Even before this deployment it was clear that the E-2 was a vast improvement over the E-1, but also that its radar and avionics were not yet up to required specifications. Foremost the 'fixed-program type' CP-588 computer of the E-2A could simply not cope with the complexity of the mission. Early operations from the Dixie and Yankee Stations highlighted additional problems with the aircraft: Tracer operators had heard 'scuttlebutt' going around the AEW community that one of the new Hawkeyes had its rear fuselage and tail section pulled off in an arrested landing. Another had reportedly suffered a total hydraulics failure when one of the new fibreglass propellers disintegrated in mid-air and caused a discarded blade to penetrate the fuselage skin and sever the primary hydraulics lines. That was bad enough but the prop debris also severed the lines feeding the secondary system. All of this created distrust in the new aircraft at a very vulnerable time in its development.

The US Navy required a number of changes and improvements to the early Hawkeye, including some major work on reprogramming the CP-588, but in those days even

An RVAW-110 'Firebirds' E-2B seen at NAS Miramar, California in October 1979. (Richard Vandervord)

minor software upgrades proved incredibly time consuming, and often necessitated hardware changes to the drum memory. Eventually, the Airborne Tactical Data System (ATDS) Project Office found no other solution but to entirely replace the original processor with the new Litton L-304F computer, whose dual-core magnetic memory could hold up to 80,000 commands of 32-bit RAM memory, and had a memory cycle of only 2.2 milliseconds. Further improvements addressed the inadequate cooling system for the avionics and computer and various other problems, but before these could be fixed, Grumman was forced to stop production after only 60 E-2A Hawkeyes had been rolled out, in early 1965. It was not until 20 February 1969 that a redesigned E-2A was flown for the first time. By then, the amount of improvements was sufficient to prompt its re-designation as the E-2B. A total of 49 E-2As were subsequently upgraded to the new standard, but they did not reach the fleet until November 1970, when VAW-114 was embarked aboard USS *Kitty Hawk*.

Big Eye and College Eye

Contrary to US Navy policy, the USAF hierarchy in Southeast Asia (SEA) did not initially foresee a requirement for an AEW capability. Unsurprisingly, this was to change before long, primarily because, in response to US air strikes, the Chinese and the Soviets began providing North Vietnam with advanced weaponry. Ironically, due to Washington's obsession with avoiding an escalation of the war, North Vietnam north of the Demilitarized Zone (DMZ) was left entirely undisturbed for months, free not only to acquire new equipment, but also to train the necessary personnel and develop a fully integrated air defence system (IADS) right under the nose of US assets deployed in theatre. On the contrary, when the North Vietnamese IADS was used in action for the first time, the US was caught by surprise: in early April 1965, at least one US Navy and two USAF aircraft were shot down by North Vietnamese MiG-17s. Finally, the USAF high command realised that there was a necessity for AEW aircraft, and consequently the decision was taken to deploy EC-121Ds to SEA.

At that time, the USAF had two wings operating this type: the 551st Airborne Early Warning and Control Wing (AEWCW) at Otis Air Force Base (AFB), Massachusetts, and the 552nd AEWCW stationed at McClellan AFB, California. Some aircraft had been upgraded with the installation of the AN/APS-95 radar that had an improved range, was easier to tune and less susceptible to sea clutter. Some had also received the newer AN/APX-103 height-finder radar replacing the old AN/APS-45. At this point it is worth noting that these aircraft and their crews had only ever operated within ADC, and in the defence of the CONUS: they had never exercised or operated with tactical fighter-bombers, or had any experience of how to control complex air combat between interceptors. The communications equipment in the aircraft was limited in scope and capability: indeed, it was tailored for communication with ground stations in NORAD. The radios proved too short-ranged for the new task and the radio frequencies in use in SEA were already heavily utilised for formation instructions, MiG and surface-to-air missiles (SAM) warnings, post-strike debriefs, SAR coordination and tanker hook-ups and there was no clarity over where in this complex scheme the warnings of enemy interceptors would fit in. Although each controller console was equipped with IFF interrogator kit, this was of little use, because the majority of US fighter pilots flying

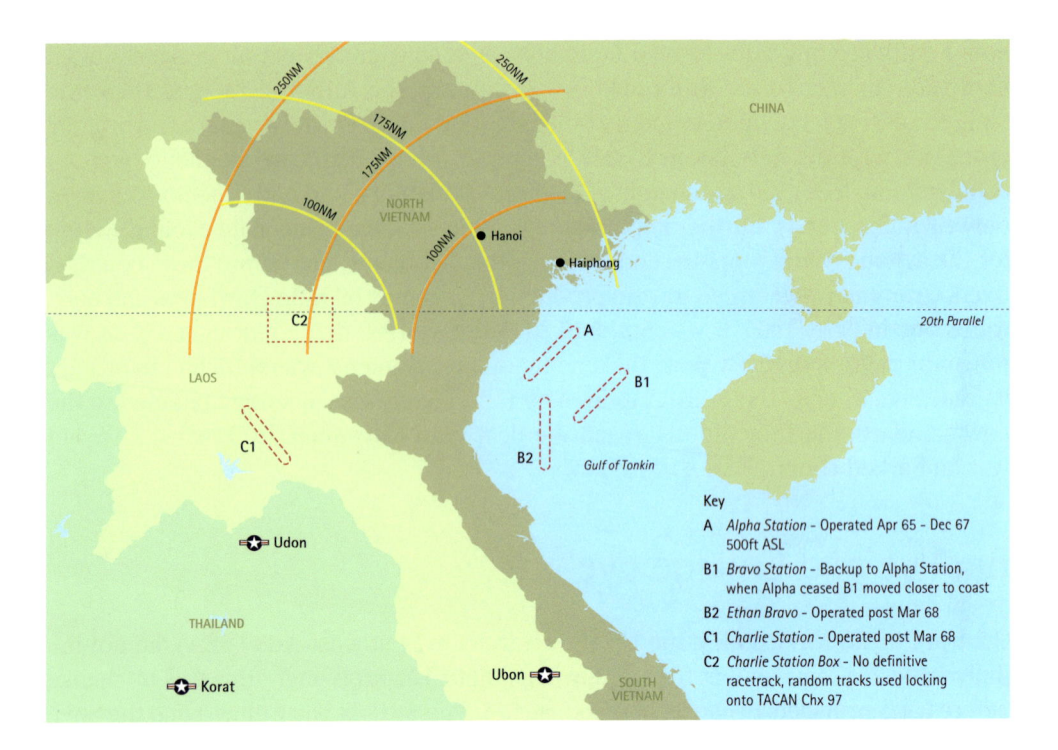

Beyond the Horizon – The History of Airborne Early Warning

A map showing the Big Eye and College Eye orbits operated from April 1965 to June 1968. (James Lawrence)

Key
- **A** *Alpha Station* – Operated Apr 65 – Dec 67 500ft ASL
- **B1** *Bravo Station* – Backup to Alpha Station, when Alpha ceased B1 moved closer to coast
- **B2** *Ethan Bravo* – Operated post Mar 68
- **C1** *Charlie Station* – Operated post Mar 68
- **C2** *Charlie Station Box* – No definitive racetrack, random tracks used locking onto TACAN Chx 97

552nd Air Control Wing (552nd ACW), US Air Force.

over the North turned their IFF transponders off: they were suspicious of the Soviets cracking their IFF codes and passing them on to the North Vietnamese GCI controllers. At that time the EC-121Ds were also an unknown quantity in regards to combined operations so far away from the CONUS, especially under hot and humid conditions.

Five aircraft and their crews from the 552nd AEWCW were grouped into a detachment designated the Big Eye Task Force (BETF) and forward deployed to Tan Son Nhut AFB, outside Saigon, in South Vietnam. The first problem the BETF crews faced was how to convince anybody else in theatre that they might be of any use, and this issue was further aggravated by the sensitive nature of the information they were collecting, which the US intelligence community insisted remained top secret. In summary, despite the BETF deployment, the North Vietnamese received additional weeks and months to further develop and improve their air defence capabilities. By 1966, the North Vietnamese ground controllers improved their skills to a level where they began attempting combined deployments of manned interceptors and Soviet-made S-75 (ASCC codename SA-2 Guideline) SAMs. From the US point of view, the North Vietnamese never became proficient in combining these two assets together simultaneously, resulting in returning US aircrews describing their operations in the North as either 'MiG days' or 'SAM days', when one or the other would be effective against their attacks. However, closer examination of North Vietnamese reports that became available after the war contradicted this, as did the evidence of US aerial loss rates.

Two EC-121Ds of the BETF flew their first sorties over the Gulf of Tonkin completely independently from the US Navy, on 16 April 1965. Very soon, their crews had to conclude that the clutter caused by the terrain inside the North, especially northwest of Hanoi, badly degraded their radar picture. A solution was found on the basis of experiences from operations off Cuba: the crews decided to fly low enough to 'bounce' emissions from their search radars off the sea surface. The returning echoes were clut-

USAF EC-121D serial number 55-0127 is accompanied by two Lockheed F-104A Starfighters during a BETF mission. (US Air Force)

ter free and provided coverage of aerial activity at medium altitude over central North Vietnam.

Initially, the BETF's Warning Stars flew on two different orbits: Ethan Alpha, at around 15-91m (50-300ft) above the sea surface some 48km (30 miles) off the coast of Haiphong, and Ethan Bravo, flown at 3,500m (10,000ft) and situated further south. The Ethan Alpha orbit proved anything but popular with crews as the aircraft's air conditioning was unable to cope with the heat and humidity at such low altitudes. With all the avionics packed into the Warning Start also generating considerable heat, several crewmen suffered heat exhaustion during the long missions flown there.

Dean Boys provided the following recollection by George Merryman, one of the pilots deployed with the BETF to South Vietnam in 1965. Merryman went on to fly some 160 combat sorties and accumulate 875.5 hours during his tour of duty in SEA:

'We transferred our main forward base to Taiwan, and from there would deploy for a week to South Vietnam. This became a routine: a week in Vietnam and a week in Taiwan. At Tan Son Nhut AFB we had three EC-121s and a small building as our forward operations centre.

'Our normal mission began with a climb to between 6,000 and 8,000ft [1,829-2,438m], to clear the mountains and proceed on an easterly course. Once over the South Vietnamese coast, we would drop down to 100ft [30m] and fly north along the coast, only far enough to remain outside the range of North Vietnamese AAA. Then we would land at Da Nang to refuel, take-off again, climb and head northeast to clear Monkey Mountain before heading over the ocean. Close to our station we would descend to 50ft [15m] above the sea, thus remaining outside the scope of North Vietnamese (ground-based) radars while utilising our systems to detect North Vietnamese MiGs and thus provide protection for our fighter-bombers heading towards their targets.

'We had a pair of [Lockheed] F-104 Starfighters under our control, held in a holding pattern: if our radar detected enemy aircraft, were would direct them to inter-

cept... Usually, we remained on station for approximately five hours, until reaching minimum fuel necessary to return to Da Nang. We would land there to refuel again, and then return to Tan Son Nhut.'²

Don E. Born was another of those flying EC-121Ds with the BETF, and described a typical Big Eye mission:

'During my first tour in Vietnam, in 1966, we flew our missions at 50ft [15m] above the water. This was for several reasons: we would be below enemy radar; we would be below the SAM envelope; and it would give a little added protection from the high-flying MiGs. It also had many disadvantages, however: at 200kt [370km/h] a twitch of the hand, a hiccup, or a gust of wind, could dump you into the drink before you could say 'which way is up'. Therefore when we would set up on station at 50ft, both pilots would have to remain in their seats at all times. The plane was flown on autopilot because it was more reliable and steady and it could hold altitude better than its human pilots. It also had quicker reaction time, especially in rough air. Even though we flew on autopilot, one pilot had to rest both hands on the control wheel with his forefinger covering the autopilot cut-off switch in case it malfunctioned.

'Another disadvantage was that the Gulf was full of Navy ships and when flying at 50ft, we literally had to climb to get over them. Many times I felt like we were making a broadside torpedo run on them. I would imagine that a good many sailors had the scare of their lives seeing this monster of an airplane coming straight for them and flying below their bridge. It was one thing to pop up and over a ship in clear weather, but the Gulf was always filled with rainsqualls and trying to maintain 50ft of altitude and 'looking-up' for ships in pouring rain is something else. Not only did we fly over Navy ships, but we also flew over foreign freighters and fleets of fishing junks. I'm sure that we probably even capsized a few junks with our prop wash at our low altitude.

'Flying these missions was very strenuous because we were confined to our seats for periods of 10-16 hours. We did not dare leave our seats even for a restroom break and the temperature in the cockpit would climb to over 100° Fahrenheit [38° Celsius]. By the time we pulled off station, we could literally wring water (of one type or another) out of our flight suits.

'These low-altitude missions were also very hard on the maintenance crews. The aircraft would return from a mission completely caked with salt residue and would have to be washed down from nose to tail. The salt spray was not only corrosive to the fuselage, but it was even worse on the big Wright R-3350 engines.

'During the break between the morning bombing sorties and the afternoon bombing sorties, we would fly down to Da Nang to refuel. This was always a welcome break in the action and a chance to stretch our legs. We were on the ground only long enough to refuel and back on station to be ready for the afternoon fireworks.

'I remember several times either coming into or leaving Da Nang, and the guys at the Monkey Mountain radar site would ask us to make a low pass and give them a 'Bubble Check'. Monkey Mountain was our most northern land radar site, but its range was not sufficient to give adequate coverage to the Hanoi and Haiphong Harbour area, or to the Chinese border. This was one of the main reasons why the EC-121D was given its particular mission. It was after one of these refuelling stops and on our way back up to station, that we nearly landed on one of the 7th Fleet carriers.

'During another sortie we were cruising north up the Gulf at 8,000ft [2,438m] and began our descent to station altitude. There were broken clouds that day with build-ups from our altitude down to about 1,500ft [457m]. As typical, the Navy and the Air Force did not recognise the existence of one another and therefore we normally had no radio contact with Navy ships nor did we even know where they would be on any given day. Well, on this given day as we let down through the clouds and broke out on the bottom, right there in front of us, right there on centreline and glide slope, was the biggest 7th Fleet carrier I had ever seen. We could not have made a better approach if we would have tried intentionally. We were honestly surprised and by the time we regained our composure, we were approaching the threshold. The landing signal officer was frantically trying to wave us off as if he honestly thought that we were serious about trying to land a 'Connie' on his carrier. I really don't know where he got that idea because I'm sure he could see that we didn't even carry a tailhook.

'With superior skill and ability, we initiated a go-around and managed to clear the bridge by several feet. It was shortly after this incident that we started to receive Navy shipboard frequencies and positioning reports during our morning briefings. I wouldn't be a bit surprised if this aborted carrier landing didn't have something to do with the Navy and the Air Force finally beginning to speak with one another!'[1]

Despite lots of hard work, the BETF EC-121s proved unable to detect any MiGs under way over North Vietnam below an altitude of 2,438m (8,000ft), and they usually only provided general information about MiG activity such as approximate location and heading. Their radar picture and contact reliability were insufficient to vector friendly fighters, and no useful height data could be gleaned as the contacts were outside the range of the AN/APS-103 height-finder. Before long it became clear that a large, slow-moving, non-manoeuvrable and unarmed aircraft like the Warning Star was effectively a sitting duck should any MiG pilot try to attack it. Therefore, the aircraft were always provided with a pair of F-104s as escorts. Indeed, following the loss of one Douglas EB-66 Destroyer to MiG-21s over Laos, the USAF took the MiG threat so seriously that if the Starfighters failed to turn up for whatever reason, the BETF mission would be scrubbed too.

The BETF missions continued apace throughout the rest of 1965, supporting Rolling Thunder air strikes, but something needed to be done to make the warnings issued by the BETF crews to the strike packages in North Vietnam quick and easy to assimilate in a very hostile environment. Instead, Washington made matters worse by declaring the Northern capital Hanoi off limits for US air power, and declaring the entire zone along the border with China a no-fly zone for its own airmen. Frustrated pilots of the USAF and US Navy sometimes violated these restrictions and on 10 May 1966, an F-105 pilot chased and shot down a North Vietnamese MiG-17 well inside Chinese airspace. This prompted not only vocal protests from Beijing, but also caused major political ripples in Washington, forcing the 7th Air Force to carry out an investigation and instigate a third EC-121D orbit above the Plain of Jars in Laos, designated Ethan Charlie. This orbit was set up with the specific purpose of watching for any further Chinese border incursions and reporting them up the command chain, and relations between the BETF crews and the fighter-bomber community were worsened by this new 'big brother' role. This additional assignment overloaded an already overstretched detachment, which was meanwhile experiencing serious maintenance problems. As a consequence, Ethan Charlie was only flown at best only every third day. Something was needed to prove the BETF's usefulness to the fighter-bomber crews of the 7th AF.

Reporting MiGs

One of the solutions introduced by Washington was to divide up 'packages' of North Vietnamese airspace between the USAF and the US Navy. Thus came into being the Route Packages (RP) system, of which 2, 3, 4 and 6B were allocated to the US Navy's TF-77, while the rest belonged to the 7th AF. Although at first this appeared to be a sound decision, the creation of RPs imposed severe drawbacks, because the ingress and egress routes of USAF formations became predictable, enabling the North to reposition its air defences under the routes, making them even more effective. Furthermore, the North Vietnamese Air Force began threatening not only US Navy and USAF fighter-bombers operating over the North, but also the EB-66 Destroyer aircraft of the USAF, deployed to provide powerful ECM against the enemy air defences during raids. The North quickly realised the importance of these aircraft and launched several attempts to shoot them down. With the EB-66s suffering frequent communication failures caused by emissions from their own ECM systems, the situation became so dangerous that the USAF was finally forced to withdraw these valuable platforms away from targets in the North, in turn decreasing their effectiveness.

The solution to the problem of reporting the MiGs' presence was found in February 1966. The BETF crews had by now become more proficient at finding and tracking MiGs. Unfortunately, however, their poor radios resulted in many of their calls never being heard by friendly crews. Something had to be done to improve the warnings and their reception. All US pilots operating in the battle space knew the position of Hanoi, so it was used as the main reference point and christened 'Bullseye'. Henceforth, all MiG positions were transmitted relevant to their position from Bullseye. For example, a call to the USAF strike package advancing northbound and citing, 'Bandits, Bullseye, 270 for 30, heading 180', would clearly indicate a MiG flight 30 miles west of Hanoi and heading south. Updates on the position of 'bandits' were passed via radio whenever there was a significant change in their heading, or at five-minute intervals. This procedure at least gave fighter-bomber pilots a slightly improved situational awareness. Poor radios in the EC-121s continued to be a problem, however. Therefore in September of the same year a system for re-transmitting MiG calls via specially equipped EC-135s or ground radio stations was developed. This improvisation was not notably effective, as the relayed calls from the EC-135s took time, and often resulted in warnings that came too late.

Quick Look

On 1 March 1966 the BETF was re-designated as the College Eye Task Force (CETF) and relocated to Ubon Royal Thai Air Force Base (RTAFB), where it joined a wing of USAF F-4 Phantom IIs. Not used to cooperation with AEW platforms and unsure of their potential benefits, the crews of the latter treated the CETF with suspicion and mistrust; in reality they were openly ignored and considered unnecessary. Ultimately, however, their co-location with their customers gave the CETF evangelists the opportunity to preach their doctrine to their erstwhile detractors and to find out what was needed by them to ensure that their warnings and assistance were of some value to the fighter community. Fighter pilots were flown in the EC-121s during missions and some very useful exchange of ideas resulted in mutual appreciation and evolving trust. The installation of IFF interrogators into the EC-121s proved particularly useful: thanks to this, the operators were able to track single fighters on their radar screen, and issue very specific threat warnings – of course, all providing that the fighter crew had selected to transmit its own assigned IFF code, which at that stage of the war was still anything but common. In a twist of irony, it was precisely an IFF-related matter that was to result in the biggest single leap forward in regards to AEW&C operations during the Vietnam War.

As mentioned previously, the EC-121Ds had been keeping a watchful eye on Cuba since 1962. In the course of these operations from McCoy AFB in Florida, the USAF and National Security Agency (NSA) had discovered that the IFF interrogators carried by AEW aircraft were capable of decoding transmissions from the SRO-2 IFF transponders installed on the Cuban MiG interceptors. Since the IFF sets fitted to all export MiGs were the same, they would also be able decode transmissions from North Vietnamese MiGs. The NSA then went a step further and developed equipment that was able to trigger a response from the SRO-2 even if passive mode was selected. Before long, commanders responsible for the CETF (with NSA approval) deployed a secret, correspondingly modified EC-121D to Vietnam for testing. The aircraft, aptly codenamed Quick Look, arrived at Tan Son Nhut AFB and conducted tests from December 1966 until January 1967.

Appropriately modified Warning Stars were now capable of triggering an IFF response from and thus detecting North Vietnamese MiGs as far as 324km (200 miles) away, often enough while they were still on the ground or at the latest while they were taking off and climbing to intercept approaching USAF strike packages. In fact, this modification proved so effective that the EC-121's primary search radar could be turned off, and the aircraft would still reliably pick up IFF responses from the MiGs. Furthermore, the Quick Look equipment proved extremely accurate and capable of providing a full picture of enemy tactics and aircraft movements, holding orbits, and numbers of MiGs scrambled. Essentially, it showed the Americans all the aspects of the air war over the North of which they were previously unaware. Unsurprisingly, a corresponding modification designated QRC-248 was hurriedly applied to all EC-121Ds at Tan Son Nhut by 21 May 1967, and henceforth the Ethan Bravo orbit was always flown by a 'Quick Look watcher', whenever USAF strike packages were operating over the North.

There were downsides to Quick Look, too: the NSA was desperate to prevent the Soviets from finding out about its discovery and use, and therefore the existence of QRC-248 was highly classified. Worried that the North Vietnamese and then subsequently the Soviets would become suspicious about their IFF transponders if the USAF began intercepting MiGs immediately after take-off, US intelligence dictated that QRC-248 was to be used by the CETF only when their main radars were not turned on. Furthermore, the NSA demanded that radar controllers on EC-121Ds must to wait for the GCI to trigger a response from the MiGs before they could actively start pursuing 'squawks' or pass relevant information to USAF interceptors, and then only to do

EC-121Ds parked on the ramp of Itazuke AB, Japan, in 1969. (Dean Boys)

so in the usual Bullseye-related warning format. It was only after the USAF suffered a string of losses to MiGs that the NSA finally granted authority for more aggressive deployment of QRC-248, on 21 July 1967.

EC-121D aircraft commander Capt Donald E. Born recalls a College Eye mission flown on 12 July 1967:

'Our crew was made up of myself as aircraft commander, my co-pilot, two navigators and two weapons controllers and 12 enlisted men, including two flight engineers, a radioman, two radar technicians and seven radar operators. We refuelled at Da Nang and returned to the skies over the Gulf of Tonkin to support the afternoon mission over North Vietnam. The strike force was already on target and about to return. They were all out and accounted for, when one of my crewmen called:

'Weapons Controller to Pilot, Weapons Controller to Pilot, would you come up on our frequency? We have a lot of action going on and I think you should listen in.'

'Roger, we're switching now', I responded as I reached up and twirled the knobs overhead. Almost immediately the cockpit was filled with tense, rapid-fire radio chatter.

'Break hard right, there's a MiG on your tail!'

'Where the hell did he come from?'

'Hang on, I'm coming to help!'

'There're three more at nine o'clock, they're coming straight in!'

'Thud Flight to MiG CAP: we've been jumped by four MiGs! Where are you? MiG CAP patrol, do you read?'

Silence...

'Thud Flight, this is College Eye. MiG CAP patrol is 20 miles [32km] due south. We're vectoring them up.'

'Tell 'em to put it in afterburner, College Eye!'

'Roger, Thud Flight. MiG CAP is 5 miles [8km] due south and has you in sight, good luck!'

Silence... and more silence.

'College Eye, College Eye, this is Phantom Flight. We're low on fuel and number two is crippled. Can you find us a gas station?'

'Roger, Phantom Flight. We'll call the tankers and tell them to head north, pronto. Can your crip make it home?'

'If he gets a quick hook-up...'

Pause...

'Phantom Flight: tankers at 2-3-0 degrees, at 130nm [240km]. 17,000ft [5,182m] and heading your way.'

'Roger, College Eye. We're on our way.'

Pause…

'Phantom Flight: tankers now at 190° at 10, turning south for a straight-in hook-up.'

'Tally-ho on the tankers, College Eye. Thanks for the help!'

Silence…

'College Eye. This is Helo 57. We're heading inland with two [Douglas A-1 Skyraider] escorts. We're coming in on a beeper of a downed pilot. Any enemy traffic in the area?'

'Roger, Helo 57. We're painting some traffic north of you at 50nm [93km]. We'll keep an eye on them for you.'

Pause…

'Helo 57, traffic appears to be two enemy MiGs. 0-1-5 at 35nm [65km], heading your way.'

'Roger, College Eye. We'll drop down into this haze layer. I hope it will hide us.'

'Traffic, 0-2-0 at 20.'

'Traffic, 0-2-0 at 5.'

Pause…

'Helo 57, you're all clear. Traffic has turned and heading back north.'

'Roger, College Eye, thanks for the warning! We've made contact with the downed pilot, we're going in for the hoist.'

'Roger, Helo 57, you're clear all the way. Have a good evening.'

'And the same to you, College Eye.'[3]

Rivet Top and Rivet Gym

In late July 1967, the CETF, consisting of six EC-121Ds (with two spare airframes held back in Taiwan) was re-located to Udon RTAFB in Thailand. Once they had arrived there the crews of the Warning Stars could start to demonstrate their ability to track MiGs to the co-located F-4 crews of the 8th Tactical Fighter Wing (TFW). Ideas were exchanged and a few of the fighter pilots flew on board the EC-121s to improve mutual understanding and garner respect and trust: all of a sudden, Ubon-based F-4 crews became some of biggest fans of the CETF, and began demanding to be passed targeting information in a more direct form from the AEW platforms.

The unit received welcome reinforcement on 9 August 1966, when the first aircraft modified to EC-121K Rivet Top standard deployed to SEA (the example in question was to be followed by four additional Rivet Gym EC-121Ks, which carried radio operators speaking Vietnamese and thus capable of listening to Northern radio communications in real time). The Rivet Top was fitted with IFF interrogators capable of reading Soviet SRO-1 and SRO-57 transponders, in addition to kit equipment for the detection of electromagnetic emissions from SA-2 SAMs. Correspondingly, their crew complement increased with the addition of six more operators. As such, these aircraft were a significant step forward in regards the development of a true AEW&C platform, and they proved so effective that the originally planned four-month test deployment was extended into 1969.

A few problems remained, however, primarily related to poor radios and thus the inability of operators to pass their valuable intelligence to crews of the 8th TFW flying

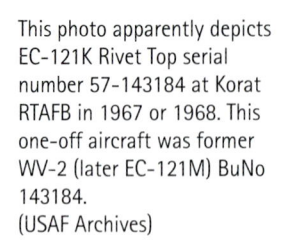

This photo apparently depicts EC-121K Rivet Top serial number 57-143184 at Korat RTAFB in 1967 or 1968. This one-off aircraft was former WV-2 (later EC-121M) BuNo 143184.
(USAF Archives)

over the North. Furthermore, Vietnamese speakers from the Rivet Gym project did not sit in front of radar screens, and therefore had no idea what USAF assets were threatened. The information they did provide could not be assimilated quickly enough to make a difference. In turn, the fact they were actively monitoring enemy frequencies had to be kept secret and thus direct warnings to 8th TFW crews were a taboo: even when permission was granted, later during the war, blindly transmitted Rivet Gym warnings were usually lost in the garble of radio traffic fired by the variety of platforms airborne over and around the North at the same time. Of course, any warning was most welcome, but it was only as good as the ability of the intended recipient to hear it. The sheer volume of radio traffic was such that many fighter-bomber crews felt overloaded and simply blocked them all out in order to concentrate on formation flying, target acquisition, and looking out for SAM launches and MiG attacks. Combined with this, the weak radios of the participating EC-121s, and lack of dedicated frequencies for AEW&C, remained the most serious problems for the rest of the air war over the North.

In October 1967, the CETF moved again, this time to Korat RTAFB, where its contingent was expanded to seven EC-121s. In early December of the same year, the Ethan Alpha orbit was suspended, as it was no longer necessary to bounce radar signals off the ocean in order to detect MiGs: the QRC-248 had meanwhile proven highly reliable in picking IFF responses even from MiGs operating at very great distances. Instead, the Ethan Bravo orbit was moved closer to the coast and slightly further south, to 20° North and 107° East. In March 1968, these orbits were re-named Ethan 01 (for the suspended Ethan Alpha) and Ethan 02, and EC-121s maintained their presence there from one hour before the morning strike package went over the North until one hour after the afternoon strike package had returned, with a quick refuel stop at Da Nang in between. The Ethan 03 orbit over the Plain of Jars was initially still used for warning crews that strayed too close to (or actually flew into) Chinese airspace, but the importance of this orbit increased following the bombing halt of 31 March 1968. From that date onwards the CETF was allocated the control of fighter-bombers, light bombers, and gunships operating along the Ho Chi Min Trail in Laos.

1 Steve Gorek, interview, via email.
2 George Merryman, interview, via email.
3 Donald E. Born, interview, via email.

COMING OF AGE, 1969–1982

After firmly establishing themselves as essential for modern air warfare during the first phase of the Vietnam War, AEW platforms experienced an ever-increasing incorporation of more sophisticated avionics for most of the 1970s and early 1980s. Their role was also gradually evolving: such aircraft were not only seen as flying radar stations providing a radar picture of the situation 'beyond the horizon', but also as a stand-alone asset able to detect potential enemies and also guide and assist friendly aircraft in most diverse ways. AEW platforms have since saved hundreds of fighters critically low on fuel by providing vectors to tanker aircraft or nearby bases, coordinated strike packages and warned them of enemy interceptors, controlled SAR missions, and monitored enemy electronic missions of different sorts.

The EC-121 provides the best example of a first-generation 'plain AEW' aircraft being converted to a second-generation AEW&C asset: in this way it matured from a 'barrier patroller' into an integrated battle-space manager. No matter how badly the EC-121 was handicapped by poor communications and poor radar pictures, and no matter how little it was trusted by the fighter pilot community it serviced, the EC-121 was slowly but surely evolving into an irreplaceable command and control platform. Similarly, the E-1B had proven itself as a good AEW workhorse, capable of simultaneously keeping watch over the fleet while assisting aircraft movement, vectoring strike packages and interceptors. Like the EC-121, the Tracer's radar performance was severely limited but its availability provided the AEW protagonists within the US Navy with a clear indication of what kind of aircraft was necessary for the future.

In similar fashion, the Gannet AEW.Mk 3 was enhanced through the addition of an AMTI upgrade and became accepted within the Fleet Air Arm as an essential multi-tasking cog in the machinery of the daily flying routine. Finally, the Soviets were easing themselves into the game through the deployment of the Tu-126, while keenly monitoring air-warfare-related developments in SEA. From the early 1970s, however, the USSR began lagging behind massively in the development of necessary technologies and never managed to close the emerging gap with the West.

Difficult times

After the end of the Rolling Thunder campaign and the bombing halt of 1968, the US began a gradual drawdown of its military from SEA. As the tempo of operations decreased, most of those that were still flown concentrated on attacking the Ho Chi Minh Trail in Laos and South Vietnam. Once result of the 'Vietnamisation' policy of new

EC-121T serial number 52-3423, in service with the 552nd Airborne Early Warning Wing, takes off from Udorn RTAFB in Thailand in late 1972. (Robert Karlosky Collection)

US President Richard Nixon was the withdrawal of CETF from Thailand in June 1970 and its re-deployment to Kwangju AB in South Korea.

Once back at their bases in CONUS, different branches of the US air arms drew different conclusions from their combat experiences over Vietnam. Among the US Navy's conclusions was that its aviators were in need of a much-improved situational awareness. As a result it intensified work on further upgrades for the E-2 Hawkeye.

Preoccupied with its strike and ground-attack mission, the USAF initially devoted much less energy to further developing the technology necessary for fighting wars like the one over North Vietnam. It went ahead with projects to miniaturise the QRC-248 enemy IFF interrogator, making it small and light enough to be fitted into around a dozen F-4D Phantom IIs of the 8th TFW under the designation AN/APX-80 Combat Tree.

Tensions in SEA increased again following the US intervention in Cambodia in 1970, prompted by the build-up of the North Vietnamese military. When North Vietnamese MiG-21s launched several attempts to intercept B-52 bombers flying over Laos, the USAF felt the need to re-establish the CETF Detachment at Korat RTAFB in Thailand and equip it with seven Warning Stars. They were to be the new EC-121T variant, which included all the previously mentioned upgrades, including Rivet Gym, Rivet Top, and QRC-248. Based alongside them at Korat were several EC-121Rs of the 553rd Reconnaissance Wing.

Despite multiple warnings from Washington and fierce US aerial attacks on their supply columns along the Ho Chi Minh Trail, in March 1972 the North Vietnamese unleashed a massive new offensive against the South, crossing the DMZ in the hope of a quick victory. Intending to bring Hanoi back to the negotiating table, President Nixon reacted by ordering US air power into action: within the following weeks, around two dozen USAF air wings and a total of six US Navy aircraft carriers deployed back to the SEA theatre. Although suffering very heavy losses, the communists insisted continued their offensive, and by mid-May 1972 Washington was left with no option but to resume air strikes against targets in the North, in an attempt to cut off the flow of supplies from China and the Soviet Union to Haiphong and Hanoi towards the south. The new aerial offensive was named Linebacker and soon turned into a no-holds-barred air war that the US military commanders had wanted to wage right from the start of their intensive involvement in Vietnam. Contrary to earlier times, however, they were now able to

Chapter 3

deploy a relatively well developed and integrated combat system against the North, controlled by advanced marks of EC-121s and protected by F-4s.

One of the major differences between Operations Rolling Thunder and Linebacker was the way the USAF deployed its MiG CAPs: supported by EC-121s using the call-sign DISCO they were allowed to roam the North Vietnamese airspace in front of an ingressing strike package, actively searching for MiGs before they could get anywhere near the US bombers. Although radio communications were still very poor and remained a major handicap, the Discos provided very effective targeting information. Furthermore, they were supported by early warning from US Navy warships equipped with long-range surveillance radars using the call-sign RED CROWN.

The improved E-2B entered service a year before the start of Operation Linebacker and only VAW-114 aboard USS *Kitty Hawk* and VAW-115 aboard USS *Midway* (CV 41) were operating the aircraft and in the area when the bombing of the North was re-started. The third US Navy carrier in the Gulf of Tonkin, USS *Hancock* (CVA 19) was still operating E-1Bs of VAW-111's Det 2.

Bob Braze served as E-1B pilot with VAW-111 during Operation Linebacker:

'The E-1B was a multi-tasking combat support aircraft, unforgiving of errors common to multi-engine aircraft, and with very marginal performance on single engine. But she was to excel in her overall reliability and earned herself the respect while supporting combat operations... Most of the time the crew comprised five, including a pilot and co-pilot, two radar operators and an enlisted flight technician.

'We were always the first to launch and the last to return to the carrier. Our primary mission was flying AEW operations over Yankee Station, about 80-100nm

Carrier Airborne Early Warning Squadron 115 (VAW-115) 'Liberty Bell', US Navy.

Aircrew from VAW-111 just prior to their deployment. Bob Braze is seen second from right in the back row.
(Bob Braze)

A formation of VAW-111 'Willy Fudds' shortly before their deployment to Vietnam. (Bob Braze)

[148-185km] *from the carrier, roughly between the coast of North Vietnam and Hainan Island. We controlled air strikes and reported any surface or aerial contacts heading towards the carrier battle group.*

'The Tracer had a simple, unsophisticated radar with no computers: everything was manual. The Bellhop data-link system didn't work well and the crews in my time were not trained to use it. We cooperated a lot with RED CROWN, which for much of 1972 was the heavy cruiser USS Chicago [CG 11], especially in regards of SAR missions and MiG surveillance, but generally we operated independently. Every naval pilot knew that if they could reach the coast there was a 'Willy Fudd' close by to help.[1] To provide this protection, we often placed ourselves in precarious and dangerous positions. The North Vietnamese would often sortie a MiG in our direction. All our missions were flown without protection of CAP fighters or any other defensive measures, so our only defence was our ability to outturn any MiG and then hope that some friendly fighter would come to aid. If all else failed, we would try to take shelter in the ever-present cloud cover.'[2]

The Tracers of VAW-110 Det 4 were the last US Navy AEW&C aircraft to operate in the Gulf of Tonkin, flying missions in the area until early 1976. Throughout their SEA deployments the E-1Bs, E-2As and E-2Bs were all to suffer from the usual ground/sea-clutter related problems, which proved unsolvable even after attempts at operating at lower altitudes than normal. As US Navy strike packages made only relatively shallow penetrations into North Vietnamese airspace, they were usually capable of making use of warnings of MiGs. US Navy AEW&C assets continued to act as navigation guides for carrier-launched packages as they went overland and would identify them again as soon as they returned 'feet wet'. They were also utilised for identifying surface vessels engaged in any kind of 'unusual activities', often at no small risk to themselves.

Another very important role of the US Navy's AEW aircraft was the coordination of SAR efforts.

One of the most effective new additions during Linebacker was the establishment of an integrated control centre, codenamed Teaball, which had been set up at Nakhon Phanom RTAFB, Thailand, in late July 1972. Teaball received several types of electronic and radar input from covert and overt sources, with the aim of creating a comprehensive image of the battle-space over North Vietnam. Inputs included information collected through SIGINT ground posts, US Navy RED CROWN ships, ELINT collected by Boeing EC-135 Combat Apple, US Navy WV-2s and CETF EC-121s. In turn, Teaball controllers passed out warnings of MiGs and SAM activity via a special KC-135 (callsign LUZON) equipped to serve as an airborne radio-relay. Luzon's warnings were in turn re-transmitted back to DISCO and Red Crown.

This excellent concept was often spoiled by the 'built-in' delay of Teaball's warnings that were broadcast on the Guard frequency of 243.0MHz. With the channel already clogged up with frantic threat warnings and emergency calls, the warnings frequently arrived too late. Furthermore, Luzon had a very poor serviceability record, and there was uncertainty over which agency was actually in charge of MiG CAPs and the vectoring of fighters at various stages of a strike mission's progress.

Nevertheless, Linebacker had effectively stopped the North Vietnamese invasion in its tracks, and inflicted extensive losses on the enemy. By early October, Hanoi appeared ready to accept all the US peace proposals, and on the 23rd of that month combat operations above the 20th Parallel were stopped again. However, negotiations between the North and the US did not involve the government of South Vietnam, and when Saigon became aware of the terms agreed, it complained so bitterly that Washington was left with little choice but to re-open negotiations. The communists exploited this opportunity to backtrack on earlier agreements and by early December all of the meetings had yet again proved fruitless. Determined not to give way, the Nixon administration then imposed an ultimatum on Hanoi to re-launch meaningful negotiations within 72 hours, or face an all-out bombardment. The government of North Vietnamese ignored the warning.

Operation Linebacker II was launched on 18 December 1972. This was a maximum, 'gloves off' air war effort, centred around attacks by large formations of B-52s on all major targets in Hanoi, Haiphong and surrounding areas. Embarked aboard USS *Ranger* (CVA 61), and heavily involved in Linebacker II, Bob Braze recalled a mission flown on 19 December 1972:

'During pre-mission briefing on 19 December we received a warning that previous-day reconnaissance revealed the presence of several Soviet-built Komar and PT-6 attack boats near islands some 75-100 miles [120-160km] north of our CVBG. They could launch anti-ship missiles and torpedoes and posed a serious threat, and thus several strike sorties were flown against them and they were considered as neutralised. Our E-1B (Modex NE015, call-sign SEABAT) launched 30 minutes ahead of the strike package and we were destined to orbit a point very close to Haiphong, where we were to control the mission.

'While the strike was approaching, we received a request to search for any surface contacts to the northeast of the TF. ENS Green, my radar operator, and me had split the duties between us: Green continued to monitor the strike package and search for enemy aircraft, while I began the search for surface contacts. A few minutes later

I picked up three surface contacts proceeding at high speed directly towards the CVBG, and reported this. The reply came that two [Vought] A-7 Corsair IIs would be diverted to attack them, and Green vectored them into the area. After approaching, the A-7 pilots reported they were unable to identify the ships due to darkness, but they confirmed they were moving in direction of the carrier battle group. I contacted our superiors on the emergency radio frequency, Navy Red, and provided an updated situation report. What followed was one of strangest missions ever given to a 'Willy Fudd'.

'The admiral in charge of Red Crown then ordered us to descend to wave-top height and visually identify the unknown fast-movers. I replied that we were unarmed, but the admiral's reply was as simple as direct, 'I know that. Good luck, gentlemen.'

'With the A-7s above us and providing cover, we descended towards the surface, our pilot rolling out at 100-200ft [30-61m] above the surface and our co-pilot LTJG Laughton would monitor our altitude and try to visually acquire the targets. The only safeguard for our survival was the warning sound from the aircraft's radar altimeter when we reached 100ft [30m] above the sea. At this altitude a small mistake would be fatal to the crew. As ENS Green vectored our 'Fudd' to the unknown targets I remained in contact with both Grey Eagle and the A-7s and provided a continual situation update. It was strangely quiet within the 'Willy Fudd' as flight officer Green continued his countdown of the miles remaining to a position overhead the contacts... Moments after Green gave the call 0.5 miles [0.8km] to target LTJG Laughton called a visual sighting. He stated the ships were running with no lights and unidentified. Suddenly the dark night was lit up like daylight. Unknown to us the A-7s flew very low over the ships and released flares. What showed up under flare light were several large warships and our 'Fudd' was in between them and below the height of their masts. LT Peden then forcibly banked the 'Fudd' over to prevent a collision with the ships. The lead A-7 pilot called to say he thought he saw tracer fire coming from the ships and wanted to engage them. The 'Fudd' started a max climb out to disengage the targets. In the moment of contact it was observed that each of the ships had a hull number, so I contacted Grey Eagle. I reported that the contacts appeared to be US Navy destroyers and gave the hull numbers to Alpha Whisky. Seconds later we received orders to disengage from the targets and take no further action.

'We resumed our station and continued with our original mission. It was only after return to the USS Ranger *that we would learn what had transpired. The destroyer USS* Goldsborough *(DDG 20), along with several other destroyers, had engaged in a gun battle with North Vietnamese shore batteries and had received casualties and battle damage. In her haste to exit the area she had failed to alert the task force commander and had taken a course to reach the safety of the main fleet. It was then realised that if the 'Willy Fudd' had not gone down to wave-top level and identified them as friendly forces there was a high probability of our own forces treating them as hostile. The skill and courage of our 'Willy Fudd' crew had provided critical tactical information and undoubtedly save hundreds of friendly lives on that dark night...*

'The reward for that night over the Gulf of Tonkin was a memo from the admiral commanding Task Force 77.7 saying 'well done'.'

Chapter 3

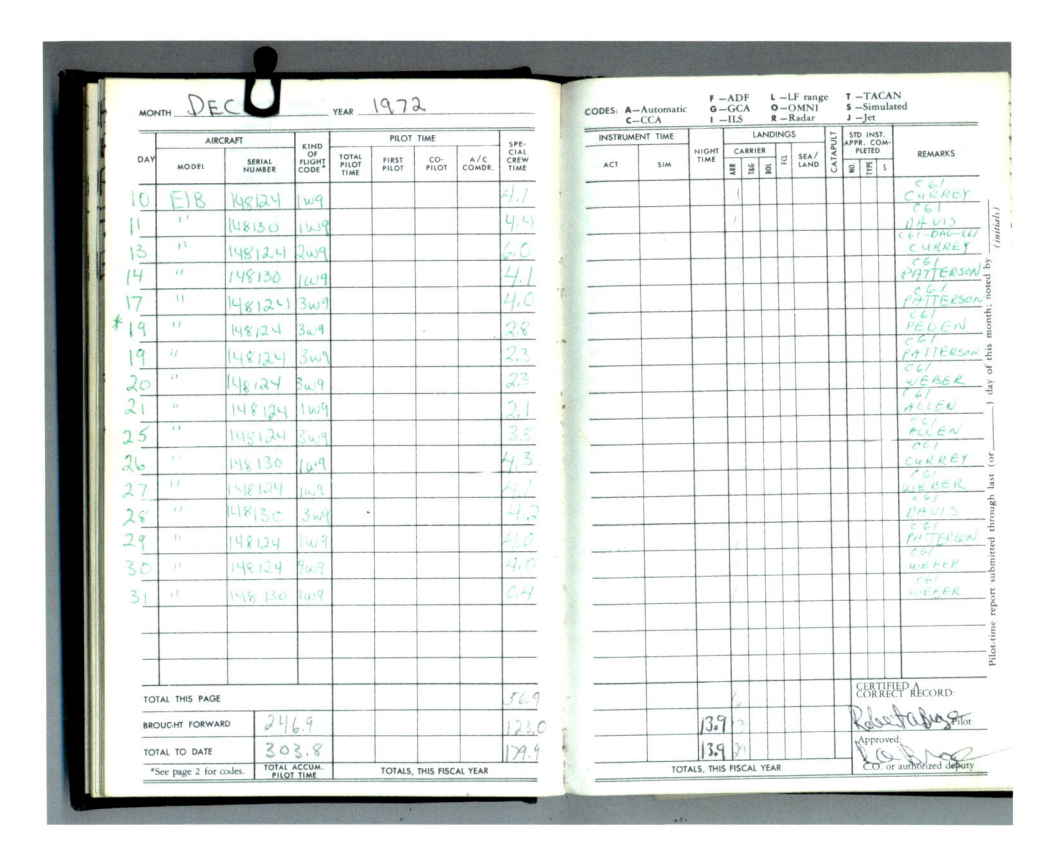

An excerpt from Braze's flight logbook showing a star annotating the mission of 19 December 1972. (Bob Braze)

Only a day later, Braze was to become involved in another dramatic mission, this time while acting as airborne strike control for a mission that ended with a SAR operation:

'Tonight the US Navy would be going to downtown Hanoi to support the US Air Force's massive bombing by B-52s operating out of Thailand. Information given to us indicated a force of over 100 heavy bombers and hundreds of support aircraft. AEW forces would stage close to the enemy coastline in support and would be standing by for search and rescue. The major responsibility for me was to search for enemy fighters. We were given reliable information that the MiGs would be airborne tonight in response to mass formation bombing by the B-52s. Navy attack aircraft would be striking anti-aircraft batteries and SAM missile launch sites.

'SEABAT would launch 45 minutes before the strike aircraft. Tonight's mission would have over 150 Navy strike aircraft from five carriers and would launch a coordinated strike designed to suppress the North Vietnamese response to the strategic bombing. The North Vietnamese forces seemed to have an endless supply of the SA-2 missiles and tonight would be no different. Once we arrived on station we began a concentrated effort to locate enemy MiGs. The strike leader checked in with SEABAT and ordered his force to go 'knockers up', meaning to arm their weapons. As they crossed over the coast the sky was filled with AAA. As the combined strike forces penetrated inland the sky filled with the exhaust trails of SA-2 missiles… The missiles were airborne and tracking targets as close as 10 miles [16km] from us and seemed concentrated on the strike group.

Beyond the Horizon – The History of Airborne Early Warning

SEABAT 015, the 'Russian' aircraft that buzzed the startled B-52 survivor on the morning of 21 December 1972. (Bob Braze)

'[Our] *radar began painting MiGs launching off the two major air bases surrounding Hanoi, and we were kept busy calling bandits to all the aircraft operating over the north that night. That's when we began hearing the sound of emergency radio beacons over our radios. This meant that aircraft had been hit and flight crews were ejecting from their aircraft. Suddenly we saw an eerie sight. Less than 10 miles away we saw something burning very brightly in the black night sky. As it approached our aircraft we identified it as a B-52, 2,000ft [610m] above us and on fire. Flames were extending back from the B-52 for a hundred feet as it descended toward the water. We then heard what seemed to be a dozen emergency beacons going off and realised that this bombers crew was bailing out. I marked off our position on my chart to give search forces a place to look for the crew. The B-52 carried a crew of 10 and a search would have to wait for first light. We spent the next hour helping many damaged and lost aircraft get back to their ships and landed after logging 2.3 hours flight time.*

05.30, 21 December 1972:

'*SEABAT, NE 015, would launch and conduct surface surveillance in the hope of finding downed airman from the battle over Hanoi last night. It was hoped that the early-morning light and clear skies would aid in any search. Arriving at the last known position of the B-52 that went down just six hours ago we started a grid search at an altitude of 2,000ft above the surface. After searching for less than 30 minutes we saw a flash of light off the port wing and turned toward it. What we saw almost immediately was a single man life raft with 1 person in it. After passing directly over it we descended to 500ft [152m] and began a turn to go back over the raft. What we saw next was the survivor jumping out of the raft and hiding beneath it. We made two more overhead passes and each time he would jump out and hide beneath his raft. Fearing he would exhaust himself and drown us discontinued our overhead passes and maintained a station up-sun of him. I called Red Crown for assistance since she had a 'Big Mother' [Sikorsky] SH-3 on board. This helicopter launched immediately and I gave it radar vectors to our position. I briefed them about the survivor and within 20 minutes they had a visual on the crewman, made a hoist pickup, and began returning to Red Crown.*

'*I asked the Big Mother crew if they had a headset they could give to the survivor so I could ask him a question. Moments later I received a call from him on the radio.*

I told him that we were the ones who found him this morning and asked if he bailed out last night when his aircraft caught fire. He said yes. I then asked why he kept jumping out of his life raft and hiding beneath it. He replied that he was afraid of a strange unidentified aircraft that was circling him and he though it might be an enemy. I then asked the Big Mother if we could come up along side of him. The helo commander gave his permission and we rendezvoused and flew lead in the formation. I then asked the survivor to look out the window and tell me what he saw. He was astounded and said it looked like a Russian plane flying next to him. I told him to look closely at the 'US NAVY' on the plane's side and said we were the aircraft that found him. He thanked us for saving his life and asked if he could do anything to thank us. I told him to send this crew a good bottle of gin and to me the signal mirror he used to flash us. He said yes.'

Linebacker II ended on 29 December 1972. It had achieved its aim of forcing the North Vietnamese back to the negotiation table, in turn enabling the final US disengagement from Vietnam. On 30 April 1975, North Vietnamese Army units entered Saigon. The US Navy was still operating off the coast, and conducted Operation Frequent Wind, a large-scale evacuation of remaining US citizens and thousands of Vietnamese from the South, during which E-1Bs of RVAW-110 Det 6 flew patrols from the USS *Oriskany* (CV 34). The last Tracers were phased out of active service in 1976, following their final cruise aboard USS *Franklin Roosevelt* (CVA 42).

Tuning the eye of the fleet

The end of the Vietnam War did not mean the end of US Navy's efforts to further improve the E-2 Hawkeye. Bob Braze, one of a few pilots that had the opportunity to fly both the earlier E-1B and the E-2B, compares them both:

'As a naval aviator I was given the opportunity to qualify in two of the Navy's AEW aircraft, the E-1B and the E-2B Hawkeye. These two aircraft were very different in both flight characteristics and varied distinctly in their airborne radar equipment. The 'Fudd' was designed as an interim replacement for the AD-5W Skyraider. The E-1B was a twin-engine radial-powered propeller-driven aircraft with limited range and performance. At a gross weight of 27,500lb [12,474kg] she could fly at a speed of 120kt [222km/h] and in Vietnam's heat and humidity could barely reach 10,000ft [3,048m] in altitude. The loss of an engine on catapult launch meant the airplane would crash into the ocean only seconds after take-off. A loss of an engine at altitude meant ditching in the sea or, if able, a landing aboard the carrier using the barricade.

'The E-2, on the other hand, was a turbine-powered propeller-driven aircraft of over 50,000lb [22,680kg] weight, a service ceiling of 35,000ft [10,668m], and a speed of 350kt [648km/h]. It normally operated on station at an altitude of 30,000ft [9,144m]. This gave it a distinct advantage over the E-1B because it had a greater radar range, excellent radio communications, and long-range HF data transmissions. The distance that long-range radar could 'paint', or acquire a target, varied by altitude. Higher altitude meant greater range.

'The NFO in the E-1B depended on his individual ability to identify a target and estimate its size, speed, and direction of flight. The radar system also had the

ability to identify friend from foe by the use of an electronic transponder interrogator. This identification friend or foe/selective identification feature (IFF/SIF) could instantaneously identify a friendly aircraft or ship. The NFO in the 'Willy Fudd' could identify MiGs flown by the North Vietnamese. This was used many times to locate the North Vietnamese MiGs and then directly control friendly fighter aircraft against them. Another very important function of the IFF/SIF system was to quickly locate and identify an aircraft in distress and guide it to safety while getting rescue forces into action. Its rugged simplicity meant that the 'Fudd' was extremely reliable and rarely missed its assigned missions.

'The E-2B Hawkeye was a totally different radar platform. It used long-range radar at an optimal 1,000,000 watts of radiated energy. Its main drawback was that it was a hydraulic nightmare for maintenance crews. Its reliability wasn't even close to the E-1Bs. The NFOs who operated the radar system were faced with a computer that detected a target then presented the contact as a computer-generated image. All images were the same size and speed and course was displayed digitally. While the operator in the E-1B used a grease pencil to track targets the E-2B operator used push buttons and an electronic graphite-tipped pen to interface with the targets displayed on its 'radar screen'.

'The Hawkeye had a marvellous flight envelope but suffered from serious problems with its composite propellers. Because of the possibility of damage to the individual blades the Hawkeye was limited to 200kt [371km/h] indicated airspeed. It also faced serious problems with the hydraulic system and several aircraft were lost due to leaks in the hydraulic flight control actuators. What I remember most was it took all five crewmen to pre-flight the aircraft for hydraulic leaks and 25 to 30 per cent of flights had to be cancelled. The E-2B's radar system was coupled to a computer that was adapted for airborne use. The system was called an Airborne Tactical Data System and could transfer all the data collected directly to other E-2 aircraft and to ships equipped with the Naval Tactical Data System via HF radio.

'The NFO in the E-1B stored his data directly on the scope using a grease pencil and on his kneeboard. Very simple and uncomplicated. As an NFO on an E-2B all data you collected was stored in the computer and if the computer 'crashed' all data

An E-2C from VAW-124 'Bear Aces' returns from a mission over the Mediterranean Sea, performing a perfect number 3 wire landing.
(PHAA Chris Thamann/ US Navy)

was lost. That was an extremely frustrating situation of which was a normal occurrence in the E-2B model. The Navy went to great lengths to train the NFOs assigned to AEW aircraft. The ability to scan, track, and identify unknown and friendly contacts and talk on five different radios at once was phenomenal. NFOs in the E-1B and E-2 aircraft became the finest airborne radar controllers in the naval task forces and earned the admiration and respect by each of the air wings and ships they served.'

Designed with development potential right from the start, the E-2 has been continuously improved ever since. Intensive testing and scores of minor changes gradually solved reliability issues and increased availability. In 1968 the US Navy ordered the next variant, the E-2C, the first prototype of which flew on 20 January 1971, and which entered service with VAW-123 two years later. The E-2C variant was equipped with the new AN/APS-120 radar with MTI, the Litton OL-77 central data processor (integrating two L-304 computers) and the AN/ALR-59 passive detection system. Much of its avionics were automated to a degree where detection and tracking would be triggered by both the OL-93/AP radar-detection system and the OL-76/AP IFF interrogator detecting the target, indicating its position, height, speed and identification to the OL-77 ASQ computer. The latter then nominated selected targets and could direct up to 20 friendly fighters using the ATDS and NTDS data. The E-2C went on to equip 20 squadrons in the US Navy fleet.

The AEW gap

While the US Navy was making progress in leaps and bounds with its AEW aircraft in the early 1970s, in the UK the government decided to impose yet another round of spending cuts. With the RAF providing guarantees for AEW and anti-shipping support for the fleet, the Royal Navy's role was reduced to what NATO required from it: primarily operations off Norway in the case of a major conflict with the Warsaw Pact. Correspondingly a decision was taken to withdraw from service all the major aircraft carriers and replace them with 'through-deck cruisers', primarily dedicated to ASW but

Seen at RAF Lossiemouth, Shackleton AEW.Mk. 2 WR960 is currently preserved at the Museum of Science and Industry in Manchester. (Mike Freer)

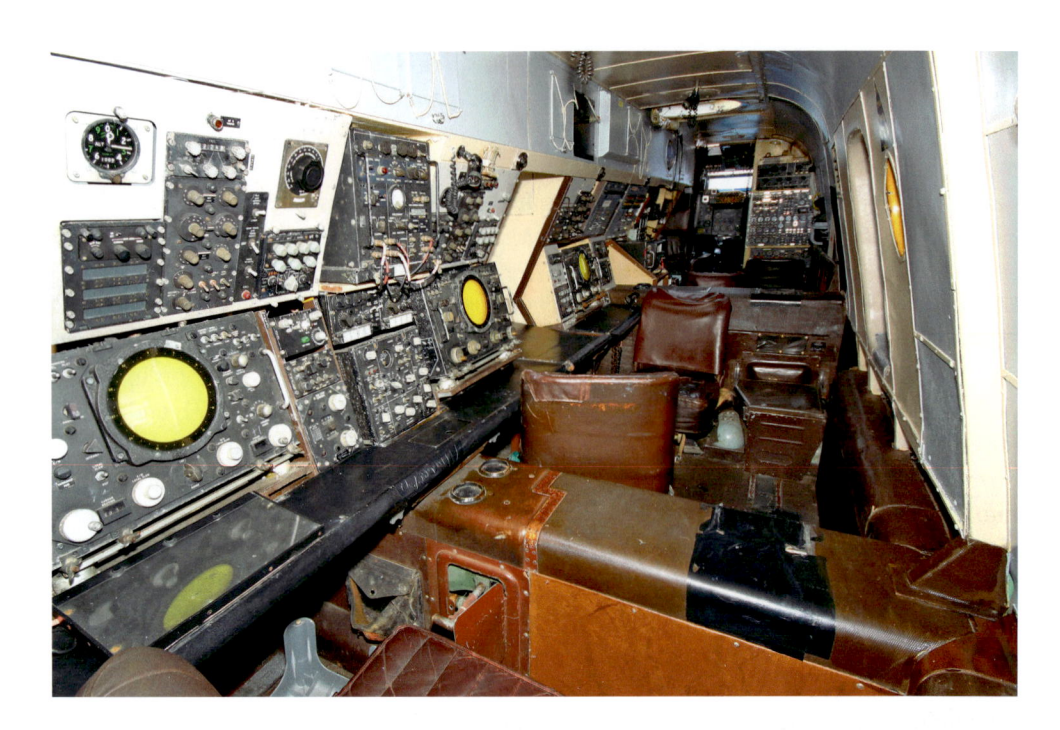

Beyond the Horizon – The History of Airborne Early Warning

Inside the fuselage of a Shackleton AEW Mk. 2, with the radar operators' positions on the port side. Note the massive wing spars that run through the cabin.
(Darren Wilson)

No. 8 Squadron, Royal Air Force.

with the ability to support vertical/short take-off and landing (V/STOL) fighters. However, there was to be no 'blue water' AEW protection for these vessels, an omission that was to cost heavily in ships and lives in the 1982 Falklands War.

British attempts to procure an indigenous AEW platform for the RAF had been subject to poor planning, nepotism and an overriding national interest from the aviation industry's trade unions. The result was that when the Gannet AEW.Mk 3 was withdrawn, the best the RAF could come up with was to equip some ancient Avro Shackleton maritime patrol aircraft with the Gannet's AN/APS-20F radars grafted under the nose. It also pilfered the Fleet Air Arm for competent radar operators to initially crew the aircraft and train the RAF operators. The Shackleton AEW.Mk 2 was only supposed to be a stopgap until the much touted and long-awaited British Aerospace Nimrod AEW.Mk 3 came on stream; in the end it would serve until 1991.

Hawker Siddeley's works at Bitteswell refurbished 12 ancient Shackleton MR.Mk 2s as AEW.Mk 2s (three additional aircraft were modified to serve for crew-training purposes). Other internal modifications related to the AN/APX-7 IFF interrogator and consoles for three radar operators, one of whom was separated from the others by the main wing spar. The converted Shackletons entered service with No. 8 Squadron that had been re-established at RAF Lossiemouth on 11 April 1972.

The crew of a Shackleton comprised two pilots, two navigators, a flight engineer and five electronic equipment specialists (including three radar operators). The aircraft was equipped with a small rest area and a galley at the extreme rear of the fuselage.

The initial plan envisaged that although operated by the RAF, the Shackleton would primarily provide AEW cover for the fleet. Initially, most electronic equipment specialists were provided by 849 NAS, who by then had developed decades of AEW experience. However, before long it dawned upon the Air Staff that there was a yawing gap in the Home Defence Region of the UK, caused by the fast new jet bombers that had

Chapter 3

Shackleton AEW Mk.2 WL747 from No. 8 Squadron at RAF Lossiemouth in May 1990. This aircraft, c/n R3696239005, is currently preserved at Paphos, Cyprus.
(Mike Freer)

meanwhile entered service in the USSR. These were able to approach the UK at low altitude and thus avoid detection by ground-based radars, launch their nuclear-tipped weapons and disappear before anybody detected them. Correspondingly, the primary role of No. 8 Squadron became air defence of the UK as part of the RAF's No. 11 Group. The RN thus found itself without any kind of AEW coverage.

No. 8 Squadron settled down at RAF Lossiemouth in August 1973, and soon had its hands full hunting all sorts of Soviet long-range reconnaissance aircraft, bombers, surface vessels, and submarines. One of pilots that joined the unit in the early days was Bill Howard, who recalls:

'I joined No. 8 Squadron in early 1972, as it was forming. The 'birth' of this unique squadron was not without its problems. The HQ 18 (Maritime) Group that operated these aircraft at earlier times were doubtless relieved finally to be rid of the 'Old Grey Lady', as they established their fleet of shiny new [Hawker Siddeley] Nimrod maritime patrol aircraft. Meanwhile HQ 11 (Fighter) Group viewed the unwelcome acquisition of a four-engined vintage antique with a mixture of disbelief and distain. The front-end crews were hairy old Coastal Command/maritime types (pilots, flight engineers and navigators) who would rather have been chasing submarines or doing a rewarding SAR job, while the rear-end chaps were a mix of RN/RAF radar operators, sorely piqued at the loss of their beloved aircraft carriers, and who were appalled at the prospect of spending up to 10 hours in a cold, noisy Shackleton. Somehow, however, we managed to get it all together.

'Ten complete crews were eventually established: the earliest were all experienced Shackleton/Gannet crews. Only later did we start our own in-house training of crews new to the type and role. Initially the tactical crew were all ex-carrier personnel, either RN or RAF exchange officers. As time progressed these were gradually replaced by all RAF guys. We did train one RN pilot, but when I left, in 1981, there was only one RN representative left, too.

'The Mk 2 tailwheel 'Shack' was a notoriously tricky beast to land for the uninitiated and even the 'old hands' were capable of 'hairy' landings from time to time. The important thing was to know how to rescue a bad landing, as failure to take timely corrective action could further exacerbate an already unsettling situation. One such incident of a mishandled landing, during a training sortie, resulted in the write-off of one of our two MR.Mk 2 training 'hacks'. The fuselage was salvaged, set on concrete plinths, and ingeniously converted into a static crew-training simula-

Beyond the Horizon – The History of Airborne Early Warning

Shackleton serial number WR960 nicknamed *Dougal*, seen in 1982.
(SSgt Coronet Cactus)

tor for the navigation/radar crew. Equipped with radar scopes and communications equipment, it simulated – with help of replayed radar tapes and electronically inserted target information – various realistic scenarios. Nicknamed 'The Dodo', it was, for its time, a very clever bit of improvisation and a valuable training facility.

'The 10 aircraft converted to the AEW configuration were all relatively low-hour airframes, although it should be borne in mind they were already around 20 years old and had been subjected to some of the harshest tropical/desert/marine environments imaginable. It became increasingly apparent that, far from being a stopgap for the AEW Nimrod, the AEW Shackleton was destined for a much longer life than at first thought. Accordingly, from around 1980 onwards, a programme of fitting new wing spars was undertaken to extend their fatigue life. This highly technical and expensive work was carried out by BAe at Bitteswell. I had the job of delivering some of them and then going back months later after the spar replacement to test-fly them for return to Lossiemouth.

'In the early days, the squadron hierarchy were mainly Fighter Command individuals with a smattering of ex-Coastal Command/maritime lower-echelon executives. Some earlier COs were Fighter Command navigators whose background of rushing about in the rear seats of [Gloster] Javelins could not have been particularly pertinent or of particular help when establishing a new role for a large-crew, large-aircraft outfit. But the new enterprise gradually took shape. Standard Operating Procedures (SOPs) were to some extent derived from existing Maritime or fighter ones where relevant, supplemented progressively by others proposed by No. 8 Squad-

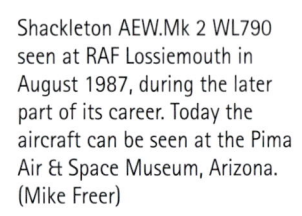

Shackleton AEW.Mk 2 WL790 seen at RAF Lossiemouth in August 1987, during the later part of its career. Today the aircraft can be seen at the Pima Air & Space Museum, Arizona.
(Mike Freer)

ron and ratified by HQ 11 Group. It was largely on a 'suck it and see' basis. The less charitable might think it was 'the blind leading the blind'. I don't recall how long it took before the squadron was officially declared ready – probably within the first year I imagine. However, when I returned to the squadron in 1977, things certainly seemed to be pretty well organised.

'We frequently cooperated with pilots of [English Electric] *Lightning interceptors. I have the very highest regard for them. I recall many a dirty winter's night in the Faeroes-Shetland gap, working with a couple of Lightnings – or sometimes a lone one – and a tanker. These single-crew chaps were roaring around at Mach plus heavens knows what, in the dark, controlling this mighty aluminium beast while peering into their 'B' scope in search of the target, responding to our directions and finally locking on to, intercepting and visually identifying the target. Then it was a mad dash to find the tanker, formate on it in the dark and take on a load of fuel for the next intercept...'*

As interceptions of Soviet aircraft and vessels began to occur at least six times a week, the Shackleton crews quickly become adept at detecting and tracking whatever the Soviets sent their way, and vectoring Lightning interceptors in return. Indeed, intensive Soviet operations over the Norwegian Sea eventually forced the unit to establish its own quick reaction alert (QRA) system, which demanded an AEW.Mk 2 be airborne within 90 minutes of the first detection of a Soviet aircraft – usually reconnaissance variants of the Tu-95 – by any of the ground-based radar stations, or other NATO assets based or operating further north. Any of the resulting 'Bear Hunter' missions would result in the award of a corresponding – yet unofficial – squadron patch. Howard continues his collection about his times with No. 8 Squadron:

'To my mind, the 'boss' who really put No. 8 Squadron on the map was Wg Cdr Phil Burton, who arrived in 1977. An experienced 'Shack' operator with a wealth of knowledge in many aspects of the RAF, he fought hard and continuously to gain recognition for his unit's vital contribution to not only Britain's air defences but also to the wider NATO commitment in that field. He was continually 'rattling 11 Group's cage' to get a fairer allocation of 'live trade' from its fighter resources, as well as pushing for frequent detachments to Keflavik and various European and Mediterranean bases to exercise with other NATO forces. His inspirational leadership, vision and dogged persistence were to be rewarded by a grateful government that slashed his squadron's strength by 40 per cent, in savage defence cuts...

'It was my sad duty to fly some of these beautiful, perfectly serviceable, now redundant 'Old Grey Ladies' to their final resting places on various airfield fire-practice dumps. Meanwhile the government blithely persisted with the AEW Nimrod variant, its development programme stumbling from crisis to crisis until the eventual, obscenely expensive cancellation of this ill-starred project (which many of us had thought should never have left the drawing board in the first place).

'We held a QRA capability at 'Lossie', with a target of getting airborne within two hours after being activated, depending on the day and the time. If it were a normal working day then the response would be much faster. However, if it was a weekend or out of hours, with crew members off base at home, then nearly two hours would be needed. We would be given a target time to arrive 'on station' and would aim to arrive at that time, there being no point in arriving too early and wasting fuel while waiting for 'trade'. Around the UK, we would work with GCI controllers

from RAF Buchan. Before going on task we would normally carry out a gentle climb, continuously recording the outside air temperature (OAT), looking particularly for evidence of an inversion, which might affect radar performance. Depending on the height of the target, we would plan to operate the barrier at an appropriate level. Normally the expected target would be high level and hence we would aim to be above any inversion layer, flying a racetrack pattern with straight legs of around 40nm [74km], generally at right angles to the expected target track. The Mk 9 autopilot was pretty basic and a bit fiddly, requiring constant adjustment. Most pilots preferred to hand-fly the aircraft. As long as you trimmed it aircraft well, it was quite responsive to hand flying. We would cruise and operate in the 5,000-10,000ft [1,524-3,048m] bracket as the aircraft was unpressurised.

'*The advent of the AEW model called for the use of engines nearly to maximum to drive the generators necessary to power the AN/APS-20F radar. This made for a noisier crew environment as the soundproofing had been removed to save weight.*

'*Although including one tactical coordinator and three or four radar operators, our teams were only providing advisory guidance to the fighters and were not allowed to control them. We had IFF kit but had no MTI in my time, although it was often talked about. The performance of the AN/APS-20F radar was variable, depending on altitude, weather, sea state and operator proficiency but the range was usually around 150nm [278km]. We operated with Lightings and, later on, with Phantoms. From time to time with other NATO aircraft, too. Usually we communicated to them on discrete frequencies on voice-UHF. We did not often have much UHF exchange with the tankers, but would maintain radar contact with them.*

'*Our normal sorties lasted around eight hours, occasionally 10-11 hours. To the ex-maritime pilots, who were used to regular 12-plus-hour sorties, this was regarded as something of a bonus! Given the antiquity of both the airframe and the radar, I think our chaps had a pretty good record in detecting targets. In the early days, the Warning Stars were still operating out of Keflavik, and their 'scopies' were pretty good. But with the advent of more modern AEW platforms, we got the impression that US operators came to rely implicitly on their computerised wizardry to analyse the information and were loath to exercise any independent judgment or initiative.*

'*As I remember, serviceability was generally good. Over my years on 'Shacks', I did experience several engine failures but I cannot recall that they were any more prevalent in later years. This was despite the fact that they were operating at much higher RPM than before. The Griffon engine was specifically designed to be at maximum efficiency at low level. On maritime duties, we seldom had cause to operate above 2,000ft [610m]. Most submarine attacks, depending on weapons used, were carried out at 100-500ft [30-152m]. SAR drops were made at 140ft [43m], ship photography at 100ft [30m].'*

The Nimrod AEW.Mk 3 – a project too far?

Through a process of evolution beginning in the early 1960s, British aviation industry scientists concluded that the best configuration for an AEW aircraft would comprise a Frequency Modulated Intermittent Continuous Wave (FMICW) radar that would be best mounted in Fore and Aft Scanner System (FASS) to give 360° coverage. Such an

Chapter 3

De Havilland Comet 4C XW626, (c/n 06419) flying out of RAE Bedford with the nose-mounted radar used for Nimrod AEW programme trials. (Mike Freer)

aircraft would have to be jet powered as the FMICW scanners would not work well on a propeller-driven aircraft. In mid-1973 several Nimrod MR.Mk 1 airframes became available as maritime patrol duties east of Suez ceased.

A de Havilland Comet 4C had been designated years before as a potential aerodynamic test bed for the forward FASS scanner and was modified to flight test at least the front part of FASS. It eventually took to the air for trials in March 1977. Wg Cdr (ret.) Ken Walne was a navigator in the RAF at the time and was detached to Royal Signals an Radar Establishment Malvern in 1979 to assist in radar trials on the AEW Comet that was flying from BAe's aerodrome at Woodford. He was also the first member of the RAF to fly on a USAF E-3A Sentry, as early as 1979, as part of an interoperability trial. It is thus apparent that the hierarchy of the RAF was keeping a foot in both camps and watching closely what was going on with both aircraft.

The wrangling between the Ministry of Defence (Procurement Executive), industry and the Air Staff continued through the early and mid-1970s with many different proposals, amendments and clarifications of the Air Staff Requirement (ASR) 400 that had laid down the specifications of the long-awaited British AEW aircraft. In retrospect it is obvious that no one agency had an oversight of the project and it appears that no one really knew what was going on at any particular time, especially the Minister of State for Defence. As flight-testing progressed it was becoming apparent that certain chickens were now coming home to roost.

Nimrod AEW.Mk 3 XZ285 (c/n 8047) was with the Aircraft & Armament Experimental Establishment (A&AEE) in 1985. (Mike Freer)

Nimrod AEW Mk. 3 XZ286 shows its planform view from below.
(Ken Walne)

Matters then came to a head. The Royal Aircraft Establishment Boscombe Down coordinated Controller Aircraft Release trials that were held between May and December 1983, and were conducted by Walne and six colleagues flying the Nimrods out of Woodford. They indicated that the aircraft was not yet ready to be accepted into service with the RAF, and the service flatly refused to accept a system that it saw as not even achieving the basic ASR 400 specification, let alone improving on it. The original specification laid down in ASR 400 and its Revision 1 by the Air Staff required that the Marconi Systems radar would be used primarily over the ocean to provide AEW of Warsaw Pact aircraft trying to penetrate the UK Air Defence Region (UKADR) at speed and low level across the North Sea. Consequently, when operating near the coast the radar picked up road vehicles, which the central processor unit then tried to assign as a potential target and displayed the moving vehicles on the operator displays. The result was that the GEC 4080M computer was working at full capacity and the radar picture was a complete clutter over land. Real problems lay in the inability of the system once it had picked up a fast-moving track to hold on to that track. That basic function was proving difficult to achieve. By 1985, the aircraft was more that three years late coming into service and the whole contract was way over the original projected budget. The politicians were now starting to sit up and take notice, as were the media. The best part of GBP1 billion had been spent and a useable system still had not been produced.

As of January 1986 the aircraft was four years late and the Shackletons were almost worn out. An RAF Joint Trials Unit (JTU) was established at RAF Waddington to see if the situation could be resolved in a military environment, with RAF crewmembers assisting in tackling the remaining problems. Walne had by now been posted as OC Nimrod AEW Software team at Waddington and although the actual mission systems avionics worked and presented no serious problems, the GEC 4080M was experiencing serious overload problems, mainly because of its lack on on-board and data-bus memory (which amounted to a paltry total of 2 megabytes). GEC promised that a new computer was on its way but this would involve more money and time. Eventually the government and the Secretary of State for Defence lost patience and in early 1986 told the contractors they had six months to get the Nimrod AEW working and that the contractors would have to pay half of the future incurred costs themselves. Concurrently, the MoD put out a request to other players in the aviation industry to submit their cost proposals for up to 11 AEW aircraft to be fully operational by 1991. That autumn it was arranged for one of the JTU Nimrods to take part in a series of fly-offs against a NATO E-3A over the North Sea. Unsurprisingly, the Nimrod performed poorly against the E-3A.

The project was officially cancelled on 18 December 1986 and the government decided that the E-3A would be purchased instead. The primary reason that the project was doomed to failure was the original ASR 400, which stated that the radar should be able to detect aircraft and helicopters with a very low 20kt Doppler velocity. This velocity meant that the PRF of the radar had to be 'medium' and not 'high'. ASR 400 was written by MoD officers and officials who envisaged that the AEW Nimrod would be operating primarily over the sea looking for hostile aircraft approaching the UK. It is questionable whether those same officials fully understood the potentially unnecessary challenges they were laying down. In the conceived context, low Doppler velocity radar was a good principle. However, when the Nimrod AEW was flying or looking over land, much of the road traffic was above the Doppler velocity and the returns saturated the data-handling system. The 80kt Doppler velocity mentioned for the E-3 radar was much higher than that of the AEW Nimrod and as such it never suffered the same overload problems. As a result of this debacle the RAF and the RN experienced an 'AEW gap' that lasted from 1973 until 1991. Indeed, the RAF's fleet of Shackletons was decreased further in 1981, when another volley of defence cuts dictated the withdrawal from service of six from 12 operational AEW.Mk 2s.

Sentry: AWACS epitomised

The US had learned significant lessons regarding AEW in the SEA conflict, and in the 1970s these lessons came together with major advances in AEW technology to contribute to the design and development of the Boeing E-3 Sentry. In the early 1970s the USAF realised that the ageing AN/APS-20B installed in the EC-121 was unable to effectively detect and track low-flying targets, especially in overland operations. This realisation came at a time both NATO and the Warsaw Pact had fielded a number of very advanced ground-based radars and weapons capable of detecting, tracking and engaging aircraft flying at altitudes above 18,000m (60,000ft). After the loss of Gary Powers' Lockheed U-2 spyplane to a Soviet SA-2 over Sverdlovsk, on 1 May 1960, low-altitude flight profiles became the order of the day. In turn, this made the capability to detect low-flying aircraft a 'must' for any new AEW platform. Such capability was already in the mind of the developers of the AN/APS-82 radar installed in the E-1B. However, that system lacked the support of computers capable of filtering out the ground clutter from actual targets. Clearly, more comprehensive research was needed to solve this problem.

In 1958 the US Navy had commissioned a programme to perfect a long-range air defence system that would comprise a carrier-borne fighter armed with long-range air-to-air missiles (AAMs). The result was the Douglas F6D Missileer that was armed with AAM-N-10 Eagle missiles guided by the huge Westinghouse AN/APQ-81 radar. The latter was capable of distinguishing changes in the Doppler frequency of moving targets at low altitude. Although the Missileer was eventually cancelled, the development of the Eagle missiles and the AN/APQ-81 was continued and after many modifications they eventually found their way into the Grumman F-14A Tomcat interceptor, in the form of AIM-54 Phoenix missiles and the AN/AWG-9 radar, respectively. More importantly, the radar research completed by Westinghouse on the Missileer programme subsequently became the basis for the AN/APY-1 radar for the E-3.

Parallel to these projects, the USAF had to solve another issue: its Tactical Air Command and the Air Defense Command were working on the development of a new AEW&C platform, but while the ADC demanded a 'simple' airborne radar system, TAC required an airborne tactical command and control system that would supervise its combat operations from above. In January 1963, experts from both commands gathered and came to an agreement to ally these roles in a single airframe. This was to be equipped with a new radar system superior to anything previously available, enabling it to search, detect, track and identify enemy and friendly aircraft, and direct weapons systems against them, and to be capable of tactical reconnaissance, alert coverage, and extended command and control in ground-based tactical areas where screening by ground units was impossible. Six months later, six companies capable of manufactur-

The first prototype Boeing EC-137D (later E-3A), serial number 71-11407, seen on rollout. (Sérgio Santana Collection)

960th Airborne Air Control Squadron (960th AACS) 'Vikings Warriors', US Air Force.

961st Airborne Air Control Squadron (961st AACS) 'Eyes of the Pacific', US Air Force.

963rd Airborne Air Control Squadron (963rd AACS) 'Blue Knights', US Air Force.

ing such a system, including General Electric, Hughes, Raytheon and Westinghouse, provided their project proposals, and the Pentagon then requested similar proposals from aircraft manufacturers that could provide a suitable carrier platform. Lockheed offered a modified EC-121 and the C-141 StarLifter transport, Douglas the DC-8 airliner, and Boeing the 707-320 airliner.

Concurrently with these processes, the USAF's Scientific Advisory Board began to research the feasibility of an airborne radar capable of scanning vertically downwards. Quickly concluding that there was no operational system with such capability, it set about to develop a system of this kind and finally overcome the problem of ground clutter. The result of this effort was the Overland Radar Technology (ORT) programme, launched in early 1965, and the selection of three companies, Hughes, Raytheon and Westinghouse, with the aim of them delivering the necessary technology. Raytheon's product was quickly dropped, while those provided by Hughes and Westinghouse entered evaluations. However, before the testing was complete, a new set of specifications was issued, demanding that the winning system had a resolution of at least one mile, be capable of tracking targets moving at speeds up to Mach 4.5, and be capable of transmitting its radar picture to other platforms and ground stations, plus the ability to provide on-board control and communication.

When all these sets of requirements were joined into one project, the selection of avionics and aircraft quickly narrowed down to the aircraft manufactured by Boeing and Douglas, of which the DC-8 was eliminated from the competition in July 1970. Boeing was thus left as the sole contender, and a pair of 707-320Bs originally constructed for use as commercial airliners were returned to the assembly line in Seattle, where they underwent major reworking aimed at reinforcing their construction and the addition of a 5,352kg (11,800lb) rotodome, which was to accommodate the AN/APY-1. The resulting aircraft received the designation EC-137D. The rotodome was an evolution of a concept pioneered by German engineer Rüdiger Kosin. In 1944, Kosin had attempted to develop an AEW version of the Arado Ar 234 jet bomber. After the war he was brought to the US under Project Paperclip, to help develop US military technology.

The vertical supports for the rotodome added an extra 1,315kg (2,900lb) to the airframe of the EC-137D and the drag induced by the rotodome structure increased the overall drag of the aircraft by around 16 per cent during climb and descent. The two EC-137D airframes used to test the two radars were 71-11407 that tested the Hughes radar and 71-11408 that tested the Westinghouse equipment. Between May and October 1972 both aircraft flew for around 300 hours in 50 sorties using prototypes of each of the proposed systems. During the evaluation phase the Westinghouse system proved to have benefited enormously from the experience gained during their development of the Doppler radar for the F6D/F-14 programs that had employed a scan-lifting technique to determine the altitude of detected contacts.

During the testing, it was noticed that the index of reflection of the radar varies according to the chemical composition of most soils: some created false targets, whose number increased relative to the altitude the aircraft was flying. The solution was to prevent the reception of radar reflections from directly below the aircraft. The tests also demonstrated that the effectiveness of the detection system varied according to the position of the target and atmospheric interference. In addition, the radar system was adversely affected by a manufacturing defect in one of the parts responsible for rotating the rotodome, something that was to be highlighted in the early models of the

Chapter 3

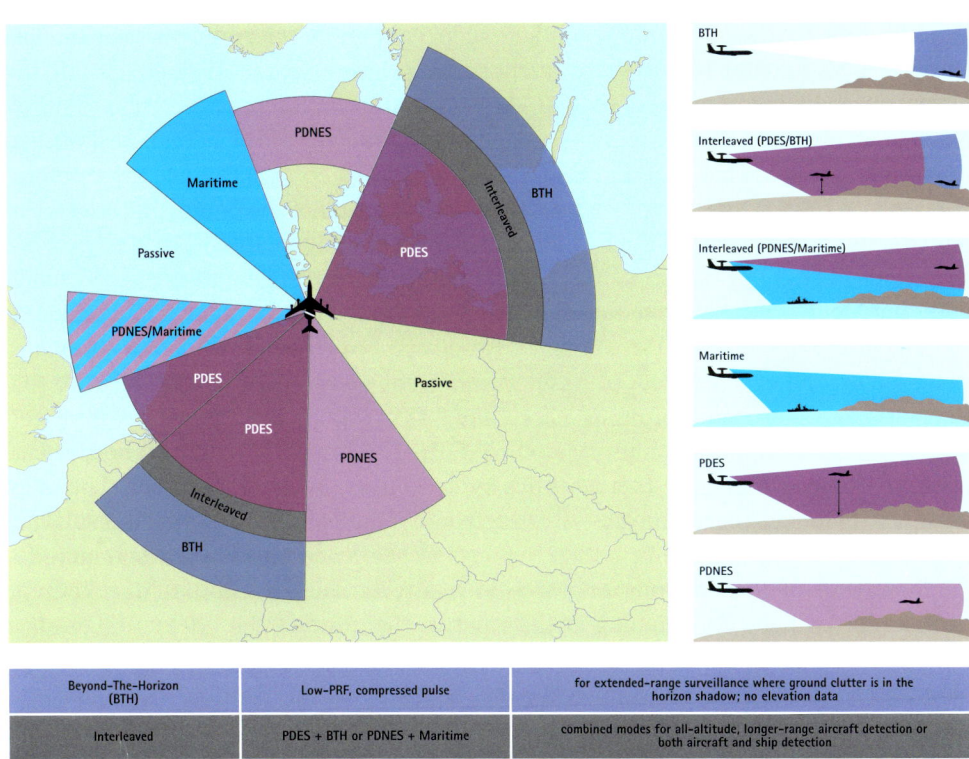

Beyond-The-Horizon (BTH)	Low-PRF, compressed pulse	for extended-range surveillance where ground clutter is in the horizon shadow; no elevation data
Interleaved	PDES + BTH or PDNES + Maritime	combined modes for all-altitude, longer-range aircraft detection or both aircraft and ship detection
Maritime	Low-PRF, compressed pulse	a very short pulse used to detect surface ships in various sea states; land masses blanked to prevent interference
Passive	No transmission	provides passive tracking of ECM sources without transmission-related vulnerability
Pulse-Doppler Elevation Scan (PDES)	High-PRF Pulse-Doppler	provides elevation data derived by an electronic vertical scan of the beam with, however, a slight loss of range
Pulse-Doppler Non-Elevation Scan (PDNES)	High-PRF Pulse-Doppler	longest-range detection of Pulse-Doppler modes; used when an increased increment of range is more important than elevation data

A plan view of the various selectable radar modes available to the E-3 mission crew and an elevation view of the various radar modes.
(James Lawrence)

E-3. The tests were concentrated purely on the detection and tracking abilities of the radars using B-52, F-4 and Convair F-106 Delta Dart aircraft as targets in environments that were entirely free from interference from ECM. Testing was carried out over varied terrains including calm seas, thick vegetation, forests and mountains.

After the evaluations the Westinghouse system proved to be the better of the two radars and beat Hughes out of the competition. The Airborne Warning and Control System (AWACS) programme now entered the Demonstration of System Integration phase, with all mission avionics and radar being tested on board airframe 71-11408 for a period of 36 hours spread across six flights in November 1972.

The location of a radar contact by airborne radar is achieved through the conjunction of three elements of the E-3's surveillance radar system, firstly the radar transmitter whose ancillary equipment is located under the cabin floor between the wing root and the rotodome pillars. The transmitter generates the radar pulse that is sent out via the antenna that is located in the rotodome above the fuselage. This slotted waveguide antenna sends out the transmitter-generated Doppler radar pulse, which hits the target aircraft. A reflected signal is then bounced back to the receiver part of the antenna, and the signal is then transferred via wiring to the radar receiver and signal processing electronics, which are located in racks inside the main fuselage. The processor part of the receiver system compares the data from successive scans, collating the contacts

964th Airborne Air Control Squadron (964th AACS) 'Phoenix', US Air Force.

965th Airborne Air Control Squadron (965th AACS) 'Falcons', US Air Force.

966th Airborne Air Control Squadron (966th AACS) 'Ravens', US Air Force.

and computing trajectories of the contacts seen, commonly referred to as 'tracks'. The Doppler aspect means that the movement of the primary contact is compared by the electronics with its orientation to the radar transmitter on the previous scan to establish its movement in relation to the scanner and not the ground. The on-board navigation systems know where the E-3 is in the sky and so, in conjunction with the processor, are able to indicate the velocity, altitude and direction of the contact after six echoes from the target aircraft. Each contact is displayed to the on-board radar operators as a digitised bright up on the 483mm (19in) multifunction display (MFD) screen, along with any other contacts picked up that are within range of the radar. The horizontal search pattern of the surveillance radar's sweep can be broken down into a maximum of 24 sectors, all of which can be assigned different search parameters and patterns. In each scan a maximum of up to 200 of any 600 possible contacts can be managed and their IFF transponders interrogated, with the IFF 'squawk' code and height data displayed 'tagged' alongside that aircraft's radar contact on the MFD screen. The overlay maps on the MFDs can be altered at each individual operator's terminal displaying geographical maps, flight information region (FIR) data and airway data as required.

The layout of the cockpit and cabin is relatively standard across the different variants of the Sentry. Beginning at the front of the aircraft, the cockpit layout has remained very similar to the original Boeing 707 airliner. It comprises two side-by-side pilot and co-pilot seats, an engineer's panel and seat on the starboard side immediately behind the co-pilot, a jump seat behind the pilot's seat and a further seat on the port side behind that. Through the bulkhead on the starboard side is a crew toilet with equipment cabinets to the rear of that either side of the fuselage. Next we come to the communications operator's workstation on the starboard side from which the VHF, UHF and SATCOM radio channels are controlled and monitored. As we proceed towards the tail of the aircraft we pass more equipment racks and cabinets and on the starboard side we come to the main computer console, again surrounded by equipment cabinets. After this we come to the main mission radar operator consoles in rows of three, orientated again to the starboard side and situated abeam the leading edge of the wing, with a gangway past them on the port side.

The first three consoles face rearwards; they are back to back with three more workstations facing forwards and a further three behind them again facing rearwards. The arrangement of which specialist crewmembers sit where varies from nation to nation and even from mission to mission, but they are all normally lead by a tactical director who will have a team of weapons specialists, fighter allocators and fighter controllers working with him or her to achieve their mission objective. As we move further rearwards, next we come to a large open area referred to by RAF Sentry crews as the 'Dance Floor'. This area leaves scope for the further expansion of consoles in future updates but is generally used as a baggage and equipment storage area during transit and positioning flights when deploying to forward operating bases. Behind the Dance Floor we come to the radar technician's station and associated console and equipment racks, abeam the trailing edge of the wing. Behind this is another area liberally populated with comms racks above which, on top of the outer fuselage, is the rotodome itself, supported on two angled, aerofoil-shaped legs whose bracings run around strengthened brackets into the keel of the aircraft. Behind the strut bracings we come to a combined passenger seating area and rest area with a least three bunks against the port side of the fuselage. Also here we find the galley with an oven and a

Chapter 3

USAF E-3A Sentry serial number 76-1605 (c/n 21435/924) seen during its arrival in Iceland on 3 May 1979, escorted by two F-4s from the USAF's 57th FIS. (Baldur Sveinsson)

A radar operator's console in an early E-3A, pictured in 1980. (Ken Walne)

hot drinks area where in-flight meals are prepared and heated for the crew on long flights. At the rear of the galley area on the starboard side and opposite the port-side rear-entry door are three more bunks with another toilet area right at the very back of the fuselage underneath the tail fin.

The E-3 has a 'double bubble' fuselage shape when viewed from the front. The lower bubble is divided into the forward and rear voids, these containing the power generators, most of the cooling equipment required to keep the multitude of consoles at optimum temperature, and the radar transmission equipment. Even though there is a vast amount of systems packed into the Sentry, from an aircrew perspective it not a cramped or unpleasant working environment – an almost unique situation for an AEW aircraft.

On 5 October 1972 the USAF announced that it was going ahead with production of the E-3 equipped with Westinghouse's AN/APY-1 radar system and that a total of 34 AWACS aircraft, which included the two development airframes, would be procured. The mass production of these airframes commenced in April 1975, and at the same time the prototype 73-1674 was sent to Ramstein AB in West Germany to conduct more tests; it flew a total of 25 sorties each averaging around four hours. During these tests the aircraft was flown over the North Sea and Adriatic to develop methods of detecting aircraft and surface vessels simultaneously. It was also flown over mountainous terrain and flew along the Inner German Border, detecting Warsaw Pact aircraft flying at speeds up to 2,141km/h (1,330mph) and an altitude of 22,680m (74,409ft).

Six months later the E-3 had been officially named Sentry and had entered service evaluation at Tinker AFB, Oklahoma, the home of the 552nd AEW&C Wing. The first Sentry to arrive there was 75-0557, in March 1977, which flew its first operational sortie during NORAD's Exercise Vigilant Overview 77, on the last day of the same month. Tests and development of the Sentry proceeded well and in April 1978 initial operational capability was declared and fully demonstrated by detachments deployed to Iceland, Japan and South Korea where it took part in Exercise Cope Jade Charlie.

Re-designated as an Airborne Warning and Control Wing (AWACW), the 552nd thus became the first unit to operate what was the most powerful AEW&C system in the world, from McClellan AFB, California, on 5 May 1976. Two months later, the unit relocated to Tinker AFB, where it came under the control of TAC, and was reorganised into two squadrons, the 963rd and 966th AWAC Squadrons. After a brief work-up, it commenced worldwide operations in early 1977, by which time two additional units, the 964th and 965th AWACS, were established with the purpose of supporting NORAD. The wing continued its expansion on 1 September 1979 with the establishment of the 960th Airborne Warning and Control Support Squadron (AWACSS) and again a month later, when the 961st AWACSS was activated (two years later both squadrons dropped 'Support' from their designation). The 552nd AWACW's first deployment occurred in October 1979, when at very short notice two E-3As and their supporting echelon were sent to South Korea immediately following the assassination of President Park Chung Hee. Finally, on 23 May 1980, the penultimate AWACS squadron was established in the form of the 961st AWACS at Kadena AB, Okinawa.

Seek Sentry: AWACS exports

The emergence of the first two fully developed AEW&C aircraft made the E-2C and E-3A very interesting for many potential customers abroad, and it was not long before both achieved export success. The first export customer for the E-3 was planned to be Iran, at that time the closest US ally in the Middle East. The Imperial Iranian Air Force (IIAF) was increased in size tenfold within a period of seven years, purchasing a wide variety of modern aircraft and construction facilities from the US throughout the 1970s. The comprehensive air defence network needed an appropriate command and control system, supported by an elaborate radar network. With Iranian topography dominated by extensive mountain ranges, even eight highly sophisticated troposcatter radar stations installed by Westinghouse were deemed insufficient: hence the IIAF insisted on obtaining E-3s.

The sale of such an advanced system caused much controversy in the US Congress and initiated protracted negotiations between Washington and Tehran. Nevertheless, the Iran prevailed, eventually signing a contract designated Project Seek Sentry, itself a part of the Project Peace Crown, which envisaged the establishment of an integrated air defence system entirely covering Iran. An order for seven E-3As was consequently accepted in 1977. Because Iran requested multiple custom-made modifications for its Sentries, nearly 50 per cent of the value of the USD1.267 billion contract was spent on the additional research and development necessary to meet the IIAF's specifications. Even today it remains unknown how many of those modifications were applied to subsequent variants and sub-variants of the E-3. The first four E-3s manufactured for Iran were still being assembled when a wave of mass demonstrations spread throughout Iran, in late 1978. Over the following weeks, this unrest turned into a revolution that ended the Shah's reign in February 1979. In the following weeks, the new government in Tehran cancelled nearly all existing military contracts, including Seek Sentry, and Washington was more than relieved to accept this and thus prevent delivery of

Chapter 3

All four IDF/AF E-2Cs saw intensive combat action during their 16 years in service. (Ofer Zidon)

extremely sensitive technology to a country that was rapidly turning into an enemy in the Middle East. The first four E-3s built for Iran were therefore re-routed, re-equipped and delivered to their next customer: the North Atlantic Treaty Organization (NATO) instead. The story of this procurement is detailed in Chapter 7.

While the E-3 proved affordable to only the best-funded export customers, the line of representatives from foreign air forces interested in the much cheaper and simpler E-2C was much longer. Officers from a number of allied nations received extensive briefings on the type during the 1970s, but the first nation to order the aircraft and to be granted the necessary export permissions was Israel.

By the time the Israel Defense Forces/Air Force (IDF/AF) began showing interest in the E-2C, Grumman and the US Navy were already in the process of introducing into service a much-improved sub-variant known as Group 0. This saw the introduction of the brand-new AN/APS-125 radar, capable now of automatically tracking ground targets. The emergence of the AN/APS-125 was closely followed by its upgrade to AN/APS-138 standard, in which the antenna restricts emissions outside of its main lobe, thus making its emissions harder to be detected by enemy sensors, supported by computer with a much-improved memory capability. About 100 E-2C Group 0s were manufactured, including four examples built for Israel. Japan was to follow with an order for 11 aircraft in 1982, Singapore with an order for four examples in 1987, and Egypt ordered five between 1987 and 1988.

No. 192 'Hawkeye' Squadron of the IDF/AF introduced its E-2Cs to service in 1978, and became the first to deploy the type in anger. On 27 June 1979, one Hawkeye controlled and vectored several formations of McDonnell Douglas F-15A Eagles and IAI Kfir C.2s into a well-staged interception of two flights of Syrian Arab Air Force (SyAAF) MiG-21s that attempted to catch a strike package en route to hit targets in Lebanon. The IDF first jammed the communications between Syrian pilots and their GCI, and then unleashed its interceptors, claiming five kills within as many minutes. Two years later, at least two E-2Cs of the IDF/AF were airborne during the Israeli attack on the construction site of two nuclear reactors outside Baghdad, in Iraq.

No. 192 'Hawkeye' Squadron, Israel Defense Forces/Air Force.

1. The 'Willy Fudd' nickname for the E-1 was derived from its original WF designation, Bob Braze
2. Braze, interview, 5 June 2013.

Chapter 4

CRUCIAL YEARS, 1980–1990

The 1980s saw a number of 'minor' conflicts occur in various hot spots, and nearly all of them either proved the worth of available AEW platforms, or made it clear beyond doubt that any actor waging modern-day aerial warfare without them was likely to suffer severe blows.

Secrets of ELF-1

Considering what happened to the Iranian order for E-3s, it might appear ironic that the Iranian Revolution indirectly led to one of first major operational deployments of USAF E-3s. Eager to establish himself as a new leader of pan-Arabist movement, and emboldened by intelligence reports stressing the decaying condition of the previously superior Iranian military, the Iraqi dictator Saddam Hussein exploited the opportunity to launch an invasion of its much larger neighbour on 22 September 1980. To his and most foreign observers' disbelief, the Iranian air arm, now re-designated as the Islamic Republic of Iran Air Force (IRIAF) hit back in strength, bombing targets all over Iraq, and stopping the Iraqi ground advance cold.

Concerned by these developments, the government of Saudi Arabia felt forced to request help from the US. On 1 October 1980, four E-3As and 200 ground personnel from the 552nd AWACW were deployed to Riyadh as part of Operation European Liaison Force One (ELF-1). What was initially planned as a temporary deployment evolved into an eight-year stay in Saudi Arabia, many aspects of which remain secret until today. The inclusion of the word 'European' in the operation's designation remains unclear, as ELF-1 consisted entirely of USAF crews. The deployment flew more than 87,000 hours in over 6,000 sorties that tracked developments in the course of the Iran-Iraq War. USAF crews were later joined by US Navy E-2Cs flying from carriers deployed in the Gulf of Oman, where they were 'close enough to see' some of most important battles of that war. Together, they tracked IRIAF forays deep into Iraqi airspace and recorded dozens of air battles in the skies over the northern Persian Gulf. ELF-1 was terminated well after the Iran-Iraq War had come to an end, on 16 April 1989.

European Liaison Force One (ELF-1), CENTCOM.

On the front lines of the Cold War

On 24 February 1982, the first E-3A manufactured for NATO landed at Geilenkirchen AB, West Germany, wearing the civilian registration LX-N90442. This measure was

NATO E-3A LX-N90446 approaches Aviano AB in Italy in November 1988.
(Sergio Gava)

NATO Airborne and Early Warning Component (NAEWC).

introduced for convenience, as the aircraft did not belong to any specific NATO nation, so the member nation with no air force, Luxembourg, was chosen. The final NAEW E-3A was delivered in 1985, ahead of schedule. The crews were drawn from 15 different nations and mixed together in composite crews. As well as the NATO AWACS unit based at Geilenkirchen, there was also a training/transport squadron operating three Boeing 707-320s, which had no rotodomes, radar or surveillance equipment. These Training Cargo Aircraft (TCA) airframes operated in a support role, training flight deck crews, and moving spare parts, baggage and general cargo around in support of NATO Airborne Early Warning Force (NAEW) detachments throughout Europe. The NAEW also established forward operating bases at Konya in Turkey, Ørland in Norway, Trapani in Italy, and Aktion in Greece.

An additional part of the NAEW consisted of No. 8 Squadron RAF that still flew six remaining Shackleton AEW.Mk 2s. The two components frequently exercised with each other and with USAF E-3s during NATO exercises, developing excellent working relations, tight cooperation and a healthy sense of competition.

Hawkeyes in 'the First War on Terror'

Only a few months after the ELF-1 detachment deployed to Saudi Arabia, the US administration of Ronald Reagan ordered the initiation of a variety of military and intelligence operations against Libya. By 1980 Libya had begun to openly support several anti-Western terrorist organisations around the world. In an effort to kick back and to be seen as supporting its allies in the region, in August 1981 the US Navy commenced a Freedom of Navigation exercise in the Gulf of Sidra, in the southern Mediterranean. A secondary aim of this exercise was to deny the Libyan claim of the entire Gulf as its territorial waters, as it had decreed in 1973, but which was completely ignored by the rest of the world. In fact, the Americans had been particularly tolerant of Col Muammar Gaddafi's offensive actions, as in both 1973 and 1980 US aircraft had been shot at by Libyan aircraft in international airspace. It was not to be allowed a third time.

On 19 August 1981 the USS *Nimitz* (CVN 68) was operating in the Gulf along with her battle group. While in this area the group was extremely cautious and maintained a

CAP that was controlled and vectored by a Hawkeye as standard operating procedure (SOP). That day two F-14A Tomcats of Fighter Squadron 41 (VF-41) were on station with an E-2C of VAW-124. The F-14s were approached and engaged by two Libyan Arab Republic Air Force (LARAF) Sukhoi Su-22Ms. During the subsequent combat, the E-2C was authorised by the CIC to engage and destroy the Libyan fighters, and the Hawkeye crew in turn authorised the Tomcats to engage. One of the Sukhois fired an R-3S (AA-2 Atoll) missile at the F-14s but it was successfully evaded. Subsequently the lead F-14 fired an AIM-9 Sidewinder which hit and destroyed one of the pair of Libyan fighters, whose pilot ejected. The second Tomcat destroyed the second aircraft a few seconds later.

On the morning of 4 January 1989 another shoot down occurred when two other F-14As, from the USS *John F. Kennedy* (CV 67) controlled by an E-2C Hawkeye of VAW-126, were operating on the southern CAP south of Crete. The aircraft were around 130km (80 miles) north of the coast of Libya in an exercise with a formation of Grumman A-6 Intruders. The Tomcat crews had been briefed to be on the alert as tensions were yet again high between Libya and the US over allegations that Gaddafi was building a chemical weapons plant. At 11.50 the Hawkeye observed four LARAF MiG-23s climb out of Al Bumbaw air base near Tobruk and head north towards the F-14s. The Hawkeye passed on authorisation from the CIC that this time the Tomcats did not have to wait until fired upon to engage the MiGs. When the MiGs were 22km (14 miles) away the lead Tomcat fired an AIM-7 Sparrow missile that failed to track. Another AIM-7 was fired at 16km (10 miles), which also failed to track. The MiGs did not break off but kept coming, and the F-14A wingman fired a Sparrow that hit and destroyed one of the MiGs, its pilot ejecting into the sea. The lead then engaged the second MiG at a distance of 2.8km (1.7 miles) from the rear with a Sidewinder. Although the second MiG pilot was also seen to safely eject it appears that no attempts were made by any SAR assets to recover either of the pilots who both died.

Carrier Airborne Early Warning Squadron 124 (VAW-124) 'Bear Aces', US Navy.

The lessons of the Falklands

On 2 April 1982 an Argentinean force invaded the Falkland Islands. The British government under Prime Minister Margaret Thatcher hastily put a Task Force (TF) that began sailing south to relieve the Islands on 5 April. Included in the TF were two 'Harrier carriers', HMS *Invincible* and *Hermes* – but no AEW aircraft. The TF would have to operate without AEW cover and rely on radar picket ships just as the RN had done in World War II.

The TF took up anchorage in Falkland Sound in preparation of landing ashore at San Carlos Bay; here Argentinean Air Force IAI Daggers and Navy Douglas A-4 Skyhawks pressed home attacks on the ships at anchor while Navy Super Étendards carried out very effective Exocet missile attacks on the radar picket ships, sinking HMS *Sheffield* and HMS *Coventry*. Bombs from A-4s eventually sank HMS *Ardent* and HMS *Antelope* and several other ships were severely damaged. Thankfully for the TF, the Super Étendards did not manage to hit either of the carriers.

British forces re-took the Islands at a heavy price in men and hardware. For the British military the AEW lesson had been learnt hard yet again. The realisation that no fleet should operate in 'blue waters' without indigenous AEW coverage resonated around the world's navies. Some, like the RN, paid very rapid attention to address deficits in this area.

Whiskey to the Rescue

After the staggeringly short-sighted idea of the Shackleton AEW.Mk 2 being able to protect the RN's deep-water operations proved so wrong, the FAA scrambled to find its own rapid solution. The result was physically imposed by the size of its new ASW carriers: its new AEW platform had to be helicopter-borne. A program to convert a Westland Sea King HAS.Mk 1 ASW helicopter into an AEW platform had been initiated during the Falklands conflict, and two Sea Kings, XV650 and XV704, were modified for this purpose in just 11 weeks. An inflatable Kevlar radome was mounted on a swinging arm on the starboard side of the helicopter just to the rear of the main entrance door. Inside this 'bag' was the Thorn-EMI ARI.5930/3 Searchwater radar scanner adapted from the radar used in the Nimrod MR.Mk 2. The swing-arm, fabricated from a piece of North Sea gas pipeline, rotated upwards and to the rear through 90° as the bag deflated when the helicopter was about to land or while on the ground. The radar arm actuator system and some of the electronics were mounted on a pallet that was stowed in the rear cabin, behind consoles for two radar operators. The rest of the ancillary equipment and cooling systems were mounted in boxes and racks immediately behind the front crew access door, leaving a very narrow walkway between the cockpit and the cabin. In order to save weight, the design necessitated deployment of only one pilot. The first flight of the system took place on 23 July 1982 and the result was designated the Sea King AEW.Mk 2. Established as D Flight of 824 NAS, both helicopters and their crews were deployed aboard the brand-new aircraft carrier HMS *Illustrious* when she was rushed towards the Falklands to relieve the RN's Task Force, in August 1982.

Fortunately, there were still a few ex-Gannet observers around the RN who were able to bring their expertise to the programme. Indeed, their participation proved essential, and this can be seen in the logbook and diary of Lt Colin Richardson, who was posted to 824 NAS as part of that initial test team. The new helicopters and their equipment experienced all manner of problems, and their operators were testing and learning how to use the aircraft during air defence exercises run while the carrier was on the way towards the Falklands. Nevertheless, the Sea King AEWs rapidly became an 'organic' component of the embarked air group and were soon closely integrated within the air defence system.

The so-called 'Whiskey' version of the Sea King experienced continuous development over the following years. As eight additional aircraft became available, D Flight was renamed as 849 NAS, the traditional squadron number for RN AEW aircraft, in November 1984. In turn, the new squadron was organised into three flights, each with three helicopters, five pilots and 10 observers. In practice, the Searchwater radar did not initially prove ideally suited for its new task as it suffered from the perennial ground clutter problems over land and only had black and white displays on which the observer had to track targets with the help of wax pencils. Over the sea, too, targets would often appear within the ring of radar clutter inside which they could not be clearly seen, and thus it was often necessary for the helicopter to change its operating altitude to shrink the ring in order to bring the bogey outside the so-called 'killer ring'.

Chapter 4

The Sea King AEW.Mk 2 prototypes. The empty ASW radomes were still fitted above the fuselage, as it was deemed too expensive to remove them at this stage.
(via Colin Richardson)

'New Model Air Force'

With the world's public attention diverted by the Falklands War and Iranian offensives in the war with Iraq, in early summer 1982 Israeli political and military leaders decided to implement their long-held plans for an invasion of southern Lebanon, with the objective of expelling the Palestinian Liberation Organisation (PLO) out of that country and thus further away from Israel's borders. As part of that invasion, the IDF/AF carried out an operation that was the first to see the deployment of an entirely new air combat management system developed by the US on the basis of experiences from Vietnam War, and from the Arab-Israeli War of October 1973.

Early Israeli operations during this campaign were marked by the usual eagerness of the IDF to make use of the latest technological developments, and to find weak spots in its enemy's defence system and exploit these to its own advantage. Far from deploying its E-2Cs in a conservative fashion, as pure 'early warning aircraft' and 'air traffic controllers', they were combined with additional methods of intelligence gathering, including UAVs equipped with electro-optical sensors and ELINT/SIGINT-gathering aircraft. Furthermore, they could rely on the support of aircraft and ground stations providing powerful ECM jamming capable of severely disrupting the operation of enemy air defences. Israeli Hawkeyes were also part of an integrated combat system that included ground-based units capable of delivering direct artillery and missile strikes against enemy ground-based air defences. Finally, the IDF/AF's E-2Cs were not only to control aerial operations of strike packages sent to attack the PLO inside Lebanon, but also to provide pilots of escorting F-15s and General Dynamics

F-16 Fighting Falcons with superior situational awareness during the missions within confined Lebanese airspace.

The declared primary objective of this 'New Model Air Force' was the PLO infrastructure established in southern Lebanon, though the IDF/AF was also searching for an opportunity to destroy a network of two ground-based early warning radar stations and their associated 19 mobile SAM sites operated by the Syrian Arab Air Defence Force (SyAADF) that had taken up positions inside southern Lebanon.

The Israeli invasion of Lebanon was launched on the late morning of 6 June, and resulted in a relatively fast although by no means trouble-free advance along the coast and into the centre of the country. By the third day of this operation, Israeli ground units came into contact with Syrian Arab Army formations deployed on the southern end of the Beka'a Valley, in eastern Lebanon, while the Syrian Arab Air Force (SyAAF) began scrambling its MiG-21s and MiG-23s into the skies over Beirut in an attempt to disturb IDF/AF operations.

It was at this moment that the Israeli political leadership unleashed its air arm: during the afternoon of 9 June, the IDF/AF executed the meticulously planned and executed Operation Mole Cricket 19 against the Syrian IADS in the Beka'a Valley: in cooperation with ground forces that fired a number of missiles and shelled Syrian positions with artillery, 90 fighter-bombers deployed in several waves managed to disable 17 SAM sites in return for no loss.

When the SyAAF rushed to scramble its own interceptors to protect Syrian ground forces, it not only oversaturated its own air control system in Lebanon, but also found this under severe attack from Israeli electronic countermeasures. Cut off from their GCI, without any kind of electronic support, Syrian pilots flew 'blind and deaf' into a giant aerial trap. Aboard their Hawkeyes, IDF commanders then began picking off and destroying one Syrian formation after another, by vectoring small formations of Eagles and Fighting Falcons into carefully staged interceptions. In possession of superior situational awareness, the Israeli fighter pilots found it easy to claim nearly 30 Syrian MiGs as shot down and 17 SAM batteries destroyed at no loss for themselves.

The Syrians remained persistent. They not only repaired and returned to service at least six of their SAM sites by the next morning, but also continued scrambling their fighters and fighter-bombers into the Lebanese skies. The Israelis repeated the exercise: carefully vectored by E-2Cs, their F-15s and F-16s claimed another 20 Syrian MiGs as shot down, while F-4, Kfir, and A-4 fighter-bombers went on to knock out the repaired SAM sites, and destroy four that were in the process of deploying into the Beka'a. On 10 June 1982 the IDF/AF once again employed its interceptors in a particularly disciplined fashion, thanks to the availability of superior situational awareness, provided by Hawkeyes. By the time a temporary ceasefire was in place, on the noon of that day, the IDF/AF claimed the destruction of an estimated 79 SyAAF fighters and fighter-bombers in five days of fighting, supposedly in exchange for only one damaged F-15B. Official Syrian sources confirmed a loss of 'more than 50' of their own fighters in exchange for a handful of claims.

From the point of view of this study, this air battle was a clear result of 40 years of massive investment in research, development and testing of AEW aircraft and their related technologies, and the rigorous training of their crews. While using technology developed in the US and by closely integrating their E-2Cs with all other available assets in its armoury, Israel produced a template for the modern-day AEW concept.

Chapter 4

An IDF/AF E-2C of No. 192 Squadron based at Hatzerim displays the in-flight refuelling capability that was unique for the Israeli Hawkeye. The capability has since been adopted for the US Navy E-2D, and should become operational in 2017.
(Ofer Zidon)

Urgent Fury

In early 1983, tensions developed between the US and Grenada, a small island-state in the Caribbean, the government of which had begun to cooperate closely with Cuba, which agreed to assist in financing the construction of a new airfield on Grenada. The administration of US President Reagan concluded that the works in question were related to developing a new base designed to support communist insurgents in Central and Southern America and began issuing warnings of this 'new threat'. By mid-October 1983, the Grenadian military staged a coup that played into the hands of 'hawks' in Washington, who in turn announced their concern for the safety of around 1,000 US medical students on the island. In a matter of days, the US military geared up and launched Operation Urgent Fury, an invasion of Grenada.

The invasion was launched on the early morning of 25 October 1983, and saw the involvement of two carrier battle groups of the US Navy, Rangers and paratroopers from the 82nd Airborne Division and a plethora of support aircraft from the USAF. E-2Cs of VAW-122 and E-3As from the 552nd AEWCW provided battle-space management for most of these forces. Despite some minor mishaps and casualties primarily related to poor intelligence and unreliable communication networks, Urgent Fury proved a success. Not only were all the students rescued, but also a new government was installed, which subsequently ran free and democratic elections in December 1984.

Carrier Airborne Early Warning Squadron 122 (VAW-122 'Steeljaws', US Navy.

Just Cause

One of the last conflicts to see deployment of AEW assets during the 1980s was the US intervention in Panama, in 1989, where the US-supported government of Manuel Noriega began to bite the hand that was feeding it through double-dealing with drug

US Air Force E-3B serial number 75-0556 lands at Aviano AB, Italy during the early days of conflict in the Balkans. (Sergio Gava)

barons, before practically declaring war on the US. After four US officers were shot at and one killed, on the evening of 16 December 1989, US President George H. Bush launched Operation Just Cause.

The intervention began in the early hours of 20 December and included over 300 aircraft and 27,000 ground troops. The fighting was pitched in few places, but largely consisted of small-scale actions by US special forces and Rangers against Noriega's hard-core supporters and gunmen.

Tasked with aerial surveillance, threat warning, vector control and control of in-flight refuelling operations, the E-3As of the 552nd AEWCW operated from NAS Roosevelt Roads in Puerto Rico. They flew a number of sorties over Panama, assisting Grumman EF-111A Ravens, Lockheed AC-130s and EC-130s, Lockheed F-117A stealth fighters, and an air bridge of 120 transport aircraft needed to reinforce and re-supply the ground forces in Panama. With the fighting all but over after Noriega's arrest, on 4 January 1990, Just Cause was declared a success, although it was an unsatisfactory outcome in the view of many, leaving an obvious US puppet in power and many Panamanians disgruntled because of damage caused to their homes and possessions by US forces.

Second-generation Soviet AEW

The inability of the Tu-126 to detect low-flying aircraft prompted the Soviet Air Force and the Moscow Institute for Engineering Research Instruments to launch studies into improving corresponding capabilities only a year after the first Soviet AEW aircraft entered service. One of reasons was that they had discovered the Tu-126's propellers badly affected the radar performance. Correspondingly, Moscow decided that its next AEW aircraft must be jet powered. Additional studies were carried out during the following years, some based on the design of the Tu-154 airliner, then in 1969 the Military-Industrial Commission (VPK) directed that Sergey V. Ilyushin's design bureau should develop a new AEW platform based on its new Il-76 jet transport. The latter was still in the process of development and was not to make its first flight until 25 March 1971, but by that time the idea was frozen, and the new aircraft was designated to enter service as the A-50.

Chapter 4

A total of eight production Tu-126s were completed, together with a single prototype.
(Tupolev OKB via Yefim Gordon)

The A-50 was fitted with a rotodome 10.8m (35ft 5in) in diameter and 2m (6ft 6in) deep, installed on a pedestal above the rear fuselage. The prototype was first flown on 15 August 1977, while the development of the necessary avionics was conducted on a modified Tu-126LL test-bed. Once all the problems related to power supply and its highly complex cooling system had been solved, the avionics suite was installed in the first true A-50 airframe (former Il-76 c/n 73409243, CCCP-76641).

There were several changes made to the original design during its conversion. In order to protect the aircrew the characteristic nose glazing was replaced by a solid dialectic radome with just a pair of small side windows under the cockpit on the nose. These windows and the cockpit side windows were coated in gold to protect the crew from waves emitted by the on-board radar. The rear gunner's compartment was deleted and replaced by two rear-facing antennas, while an in-flight refuelling probe was installed on the nose. The landing gear sponsons on each side of the fuselage were extended to accommodate navigation and communications equipment, the liquid oxygen converter and the auxiliary power unit (APU). The rear cargo clamshell doors were also deleted. Tests in the wind tunnel and in flight confirmed that the aircraft's longitudinal stability was impacted by the presence of the rotodome, which caused airflow to be directed downwards and away from the horizontal stabilisers and thus starved them of lift. This prompted the designers to install a pair of strakes on the sides of the undercarriage sponsons. It was also discovered that the rotodome caused instability and oscillations when the A-50 tried to refuel from an airborne tanker, which meant even more skill than normal was required from the pilot (similar turbulence-related issues were experienced by pilots of the E-3).

The mission crew consisted of two pilots, a flight engineer, a navigator and 10 systems operators. The radar consoles were situated on the port side of the aircraft with all remaining space taken up by equipment racks full of electronics. As in the Sentry's AN/APY-1 radar, the Shmel radar carried by the A-50 utilises the pulse-Doppler system that is transmitted by a rotating head in a rotodome mounted on the pylons above the centre section of the fuselage. It differs from the US radar in that is optimised to detect air targets and maritime targets at the same time; it achieves this capability by alternating its pulses in three frequencies whose echoes are compared and processed from analogue into digital and then filtered simultaneously by four Tsc Vm computers, which measure the distance, azimuth and elevation angle of the contacts while

Beriev A-50 'Red 50' on its way to Moscow for participation in the annual air parade in May 2013. Noteworthy are the chaff/flare dispensers below the 'Bort' number on the rear fuselage. (Sergey Krivchikov)

tracking their trajectories automatically. The tracks are then displayed on the operator workstations. There is in-built redundancy in the system that enables seamless continuity even if one of the Tsc Vm computers fails. Information on the tracks includes type and whether it is friendly or hostile. This data is presented on the operators' monitors that can be set up individually to cover specific geographical regions within the radar cover. The monitors can be zoomed in and out of scale range as required by the operator. All tracks can be monitored in real time by the ground-based central command via a secure nine-channel Rubezh-1M data-link.

The A-50 is able to vector interceptor aircraft remotely using encrypted radio transmissions passing the position, altitude and speed of the target to the interceptor through the Raduga-75 link every 10-30 seconds. Data such as fuel and weapons available are also exchanged between the interceptor and the A-50 operator. The A-50 normally flies in rectangular or figure-of-eight orbits, whichever is deemed the most effective for maximising the duration on station. Patrol areas can be coordinated with other airborne A-50s and ground facilities, using VHF/UHF radios with an average range of around 350km (217 miles), HF with a range of up to 2,000km (1,243 miles) and satellite communications that can provide global coverage.

Three aircraft were used for testing and evaluation, the first of which was converted by the Beriev works in Taganrog and made its first flight on 19 December 1978. By October 1983, two additional examples ('Red 15' and 'Red 20') became available, the former of which was used for testing of the Shmel radar and the later for testing ECM systems. The test programme and evaluation process was finally completed in 1984, allowing the A-50 to be finally released for production. About 40 A-50s were manufactured through conversion of Il-76 transports during the following year, but their subsequent development was badly hampered by a lack of money. Most of the fleet was based at Siauliai, in Latvia, but this base was abandoned when the former Soviet Republic became independent in August 1991.

Chapter 4

A rare air-to-air study of A-50 'Red 50' from above, revealing details of the rotodome, antennas and the prominent strakes on the sides of the undercarriage sponsons. (Sergey Krivchikov)

Thanks to intelligence provided by Adolf Tolkachev, an electronics engineer who was working as chief designer for the Scientific Research Institute of Radar (NIIR or NII Radar, meanwhile better known as the Phazotron Design Bureau) the Soviet Union's largest developer of military radars and avionics, the US was well informed about the appearance and capabilities of the A-50 in the early 1980s. The information in question was so sensitive, though, that the CIA did not share it with NATO until 1983. Nevertheless, no A-50s were seen in public until December 1987, when an example was intercepted by a Royal Norwegian Air Force Lockheed P-3B Orion maritime patrol aircraft.

A-50s saw little more than routine peacetime operations until January 1991, when one example was deployed to monitor Allied operations against Iraq, during Operation Desert Storm, from a base near the Caspian Sea. The situation changed in December 1995, when the Russian Air Force was mobilised to support the Russian military intervention in Chechnya. The air element of the expeditionary force totalled about 200 tactical fighters and helicopters, including a few Mil Mi-9 command and control platforms. The air war began with assaults on abandoned Soviet air bases and airfields around Chechnya, where up to 275 aircraft – primarily jet trainers – were parked. Operating from Privolzhskoy air base in conjunction with MiG-31 and Su-27P interceptors, A-50s were responsible for establishing total control over Chechen airspace and preventing any form of unauthorised aerial movement. The most famous mission involving A-50s was flown on 22 April 1996, when an example apparently supported the interception of a satellite phone used by Chechen leader Dzhokhar Dudayev, and then vectored a Su-25 attack aircraft to hit and eliminate the target.

Exactly three years later A-50s were deployed again over Chechnya, when Chechen insurgents invaded the neighbouring Republic of Dagestan. This prompted a powerful response from Moscow in the form of an offensive by 10,000 troops, supported by 1,700 combat sorties by tactical aircraft and helicopters. Once again, A-50s were apparently limited to the control of airspace over the battlefield, in cooperation with Su-27Ps. Available reports indicate that they were used to track down two Mi-8 helicopters used by insurgents to transport supplies, although it remains unknown if these were intercepted while airborne or destroyed on the ground. The Russian Air Force carried out more than 4,000 sorties in the course of this Second Chechen War, although the intensity of operations by A-50s gradually declined over time.

Chapter 5

THE NEW ERA, 1991–1999

With the end of the Cold War, in 1989-1991, it appeared as if a 'Peace Dividend' would lead to a massive drawdown of offensive weapons systems including AEW platforms. While the militaries of NATO and the dissolving Warsaw Pact were rapidly reduced in size, broader expectations were to prove incredibly naive, as new conflicts emerged within a matter of months, all of them establishing that AEW aircraft were even more of a necessity than ever before. On the contrary, as the nature of different threats changed, so the existing AEW platforms had to be adapted to meet an ever-increasing number of different tasks.

End of the Shackleton

On 30 April 1990, after controlling RAF Blackburn Buccaneers while they were firing Martel guided missiles in an exercise area in Scotland, Shackleton AEW.Mk 2 WR965 called ATC at RAF Benbecula, announcing its intention to make some practice visual approaches to their runway. Nine minutes later, another call was heard, declaring that the weather was not good enough for this and that the aircraft would be departing the area. The wreckage of the Shackleton and the bodies of all 11 of its crew were found on a hillside on the Isle of Harris. Subsequent investigation indicated that the aircraft was serviceable and under power when it hit the hillside while navigating through low clouds.

This was a sad epitaph to the Shackleton's impeccable service record, all the more poignant as it happened in that type's last months of service. On 1 July 1991, No. 8 Squadron moved from RAF Lossiemouth to RAF Waddington and re-equipped with the E-3D Sentry, tailored to British requirements.

Shackleton AEW.Mk 2 WR965 *Dill* was lost in a fatal accident in April 1990. (Mike Freer)

Desert Shield

While the RAF was still gearing up for acquisition of its E-3Ds, trouble was brewing in the Middle East. Once again, Iraq' Saddam Hussein was about to launch an invasion in order to distract from political and economic problems at home. Despite weeklong tensions and concentrations of Iraqi military forces along the border, Iraq's assault on Kuwait of 2 August 1990 achieved surprise and the invasion was completed within 36 hours. The UN condemned the invasion and demanded a complete withdrawal of Iraqi forces from Kuwait, while Saudi Arabia requested Western and US help in defence of its country, on 6 August 1990. Opening Operation Desert Shield, the first US troops reached Saudi Arabia within 48 hours of the request. Over the next four months, they were followed by hundreds of aircraft, six aircraft carriers and around 200,000 soldiers from 34 different countries. On 29 November 1990, the UN Security Council passed Resolution 678, which demanded that Iraq withdraw from Kuwait by 15 January, or face 'all necessary means' that would force it to do so.

Except for the AWACS crews of No. 18 Squadron RSAF who were flying from home soil and covering the arrival of US and other aircraft, the first coalition AEW aircraft to arrive were E-3As of the 552nd AWACW, on 7 August 1990. Due to their lengthy deployment with the ELF-1, their crews were familiar with the operating environment and they divided up areas of responsibility (AORs) between them and the Saudis. Bill Richards was with the unit during the second half of 1990:

'On 6 August 1990 we departed for Riyadh… Six E-3 Sentries departed from our base in Oklahoma City, Oklahoma for the 16-hour-plus flight to the 'Sand Box'. After two in-flight refuellings we landed on 8 August. As the aircraft landed and the crews were billeted at various hotels around Riyadh we were placed into immediate crew rest.

'My crew was selected to fly the first US AWACS mission of what was to become known as Operation Desert Shield. We came out of crew rest early in the morning of 9 August and reported for duty at Riyadh AB. Our pre-mission brief was a little sparse on information regarding what assets we would be working with but our primary objective was to get AWACS airborne and our radar operational and looking at Iraq. The mission was planned to last about 16 hours, which meant that we would need an air refuelling at about halfway through the flight. We were assured that a USAF KC-135 tanker would be launched from Jeddah at the appropriate time to rendezvous and refuel us. With that information in hand we headed out to the aircraft to prepare for launch.

'The launch and climb to our orbit area went smoothly. As we went through the routine of powering up our equipment, especially the radar, we were anxiously wondering if we would be alone in the airspace over Saudi Arabia. This was cause for great concern as the E-3 is essentially a defenceless aircraft. That, plus our 'run away' speed was not too impressive and we could easily be overtaken by Iraqi fighters and shot down. Once we completed console check-out and the radar displays came to life, we were very much relieved to see four USAF F-15 Eagles of the 1st TFW flying a CAP about 100 miles [161km] north of us. In short order we had the Eagles up on our frequency and under control. The mission progressed smoothly with a few anxious moments as we investigated various unknown radar contacts with the Eagles.

'Later on, in one bizarre intercept we actually vectored the Eagles in on a low-altitude, radar-only contact that turned out to be a car egressing from Kuwait ahead of Saddam's army at a very high rate of speed! Every four hours the fighters would

Chapter 5

A USAF E-3 detachment at Riyadh, Saudi Arabia. Note the Pilatus PC-9s and old BAC Strikemasters in the background.
(via William Richards)

be replaced with a new flight. Somewhere around nine hours into the mission, as we approached the point were we would have to return to base if we were not refuelled in-flight, it became apparent that our tanker was going to be a no-show and that we would be forced to go off station and land. This caused a great deal of concern with the air staff on the ground, as there would be a break in the airborne early warning coverage while the next AWACS crew was awakened and hustled out to Riyadh. At this point it hit me that there was still a massive deployment of US forces pouring into Saudi Arabia by air and that surely there were some USAF tankers airborne that might have some fuel on board available for off-load. So I advised the flight deck and the mission crew commander of my plan and told them to prepare for either an ad hoc air refuelling rendezvous or a minimum-fuel return to base. 'With my fingers crossed, I transmitted on UHF guard (the international emergency frequency). 'Any US Air Force tankers in the area please come up on guard'. Within minutes I was talking to two KC-135s and a [McDonnell Douglas] KC-10. I quickly checked with each of them and determined that only the KC-10 had fuel available for off-load to allow us to stay on station until relived by the next E-3. The KC-10 reported that he was in the landing pattern at Riyadh. After getting radar contact on him I had him advise ATC that he was leaving the pattern and gave him an initial vector to our direction and instructions to climb to 24,000ft [7,315m]. Once he acknowledged those instructions I asked him if he was familiar with the frequency 'Winchester', as we could not remain on guard and conduct our air refuelling operation. Winchester is a nickname for UHF frequency 303.0 (or '30-30'). He said he was and I told him to push to Winchester and 'go green' (meaning to switch to the secure encrypted comms). A few moments later the KC-10 was up on frequency and headed our direction. I instructed the mission crew commander and flight deck that a tanker was headed our direction and that this was going to be quick and dirty. Everyone quickly began preparations for air refuelling and I instructed the flight deck to come up on frequency and I would brief them and the tanker on the rendezvous. 'The aircraft commander of the E-3 was a good friend of mine, Maj Larry Beam, and *the best pilot I've ever had the privilege of flying with. I knew that if anyone could make this happen it was Larry Beam. I spelled out the plan to both flight decks that I was going to vector the aircraft head-on and perform a 90-90 turn-on. A 90-90 turn-on is where both aircraft are vectored directly at each other, separated by a 1,000ft [305m] vertically, and then at the precise moment one aircraft is given a left 90° turn and the other is give a right 90° turn. If the turns are given at the correct time the receiver should roll out a mile or two astern the tanker. To further complicate this manoeuvre there were several commercial aircraft on a civil airway between the E-3 and the KC-10. So I had to keep the E-3 at 31,000ft [9,449m] and the tanker at 24,000ft [7,315m]. Normally the receiver would be 1,000ft below the tanker to facilitate the rendezvous. Doing a 90-90 turn-on is the quickest way to get two aircraft rendezvoused and it's also one of the more difficult control manoeuvres possible. 'With both aircraft briefed and the wheels set in motion I pointed the E-3 nose-to-nose at the KC-10. At about 11 miles [17.7km] separation I turned the KC-10 90° to the right and 2 miles [3.2km] later I turned the E-3 90° to the left. Within a few minutes the E-3 flight crew reported the tanker in sight and that we were rolling out 1 mile [1.6km] behind the tanker. As the commercial aircraft cleared our paths I advised the flight deck that they could begin their descent if they still had the tanker in sight. They advised that they had a visual on the tanker and were starting descent. The mission crew quickly ran through the radar shutdown procedures and we all strapped in as Larry pulled out the stops and we descended 8,000ft [2,438m] in a matter of seconds to drop in behind the tanker to begin refuelling. 'After that refuelling the rest of the mission was quite uneventful and I realised that necessity can really be the mother of invention.'*[1]

Up until the end of Desert Shield, on 15 January 1991, the USAF and RSAF AWACS crews had accumulated 5,052 flying hours in 397 sorties. The repeated pattern of sorties 24 hours a day in the same orbit locations inside Saudi airspace primarily served to 'de-sensitise' the Iraqi Air Force to the presence of AWACS aircraft nearby: at the start of the subsequent attack the Iraqi military would be caught by surprise.

Baghdad and Adnan

While Richards and other crews from the 552nd AEWCW were patrolling the Saudi skies, Iraq was still trying to obtain at least a rudimentary AEW capability. Inspired by Soviet work on converting Il-76 transports to become A-50s, Iraq pursued a similar programme, known as Baghdad. During the eight years of the war with Iran, Iraqi air defences had been continuously taken by surprise by Iranian air attacks, Iranian pilots becoming experts at low flying in order to avoid detection. Iraq was already manufacturing Tigre-G early warning radars of French design under licence as the SDA-G, and with the Iraqi Air Force (IrAF) operating more than 50 Il-76MD transports, many of them wearing the livery of Iraqi Airways, a suitable carrier platform was available to modify.

In 1987 the IrAF and the engineers of various Iraqi companies launched an effort to match the two, through installation of the Tigre-G antenna measuring 3 x 1.6m (19ft 8in x 9ft 10in) inside a radome, creating the Baghdad-1. The antenna was installed upside down and then enclosed by a plastic radome that projected

Chapter 5

The Simorgh received the serial number 5-8208 prior to its loss while operated by the 73rd Tactical Transport Squadron, Islamic Republic of Iran Air Force. Its crash was caused when the rotodome separated from its mount and hit the fin, making the aircraft uncontrollable. (via Tom Cooper)

from the lower rear fuselage, instead of the loading ramp. Support equipment and four consoles for operators were installed within the cargo cabin. The resulting radar scanned a 180° arc to the rear of the aircraft, which not only proved a very unusual arrangement for an AEW aircraft, but also resulted in a very disappointing performance. Namely, although developed as a specialised, mobile early warning radar, capable of detecting low-flying aircraft out to a range of 100km (62 miles), the Tigre-G was optimised to support AAA and SAMs. It proved useful for coastal surveillance and support of shore-based anti-shipping missiles. However, when installed in an aircraft it also proved heavily susceptible to ground clutter and provided a useless radar picture. Furthermore, some of the foreign scientists who had an opportunity to examine this project after the war concluded that if used for any length of time, the Baghdad-1 would have probably micro waved its own crew due to the lack of screening inside fuselage.

Unsurprisingly, by the time the Baghdad-1 made its first flight, Iraq had started work on its first 'true' AEW aircraft, originally named Baghdad-2, but later renamed as Adnan (later Adnan-2), in commemoration of the Iraqi Minister of Defence from the later stages of the war with Iran, who had been killed in an accident in 1989. Converted from Il-76MD c/n 0033449455, the Adnan included a fibreglass rotodome measuring 9m (29ft 6in) in diameter, mounted on struts above the rear fuselage. Although again equipped with the Tigre-G (by then re-designated as Salahaddin-G), this design proved slightly more successful, detecting targets out to a range of 120km (75 miles). However, this meagre range and the narrow six-degree-wide beam of emissions made it useless: in essence, anything flying above or below its radar beam remained undetected, which meant that the system had unacceptably wide blind spots.[2] Not only did the IrAF face an opponent equipped with F-14s toting AIM-54 missiles with an effective range of around 110km (68 miles), but the Salahaddin-G also proved practically blind at shorter ranges and continued to suffer from ground clutter and false echoes caused by the airframe and engines. As far as is known, except for two workstations for controllers, neither of the Adnans was ever equipped with any kind of additional COMINT, ELINT/SIGINT, or ESM installations.[3]

Joined by the second Adnan-2 (Il-76MD c/n 0083484542) in early 1990, the first prototype was still involved in research into improving signal processing of supporting hardware as of August of that year, when Iraq invaded Kuwait, prompting the UN to

The Iraqi Il-76 AEW conversion named Baghdad 1 (later Adnan 1) escaped from Iraq and was requisitioned by the Islamic Republic of Iran Air Force. It is seen here grounded at Tehran airport. (via Tom Cooper)

impose a total arms embargo on the country. While some further work was undertaken during the following months, when Iraq found itself on the receiving end of the US-led military intervention to liberate Kuwait, in January 1991, all three Adnans were still not yet accepted for service with the IrAF. Correspondingly, they were on the ground at Taqaddum air base during the night of 16 January 1991, where laser-guided bombs dropped by USAF F-117As knocked out one Adnan-2. The Adnan-1 and the sole surviving Adnan-2 were subsequently evacuated to Iran, where both were impounded and taken as war reparations for damage Iraq caused during its aggression in 1980.[4]

Desert Storm

During the massive build-up of Coalition forces in Saudi Arabia, in late 1990, it became obvious that there was a need for clear RoEs, operating procedures and detailed command and control of future combat operations against Iraq. In Vietnam the USAF and US Navy had divided up the airspace over South and North Vietnam into clearly defined areas labelled Route Packages (RPs). Although not the most efficient way of approaching the problem, this was the best option available at the time given the absence of complete radar cover of the battle-space and no effective joint planning staffs. At least the two air arms were de-conflicted from each other and their respective planners could use time and height separation between the known waves of different operators flying within their RP.

That said, something more cohesive and detailed was needed for this campaign, and this came in the form of the Air Tasking Order (ATO). The objective of the ATO is to provide command staffs, unit operations staffs and the aircrew participants with a minute-by-minute game plan of what is planned to happen at a particular time, place and altitude within the AOR. The high command dictates the policy and the ROEs of the operation and the special instructions (SPINS) that are to be adhered to by all participants. The process begins with a master target list that dictates in order of priority what type of strategic and tactical targets are to be hit and their priority. These priorities and strategies are normally decided at government level with advice from defence chiefs. Once this target list is produced it is then up to the theatre air chiefs and their planners to decide how best to implement the attacks on the designated targets with the knowledge of what 'in theatre' assets they have available to them. At base level a constant upward reporting process is gone through, in which numbers of serviceable aircraft and available aircrew are passed up the command chain to the Combined Air Operations Centre (CAOC) so that the available resources are a known quantity to the planners at all times. In a simplified and unclassified description the ATO is essentially the product of three days' worth of forward planning by central command staffs at the CAOC or, in the case of Desert Storm, Central Command based in Riyadh. The Joint Force Commander (JFC) appointed for Desert Shield and Desert Storm was Gen Norman Schwarzkopf, who was ultimately responsible for the application of all Coalition land, sea and air forces and their application in the conflict, and ultimately the targets to be struck. His initial aim was to reduce the most feared part of Saddam's war machine, the Republican Guard divisions, to below 50 per cent of their battle strength, then to take out the Iraqi command and control structures and air defences. He delegated the role of Joint Force Air Component Commander (JFACC) to Lt Gen Charles

Chapter 5

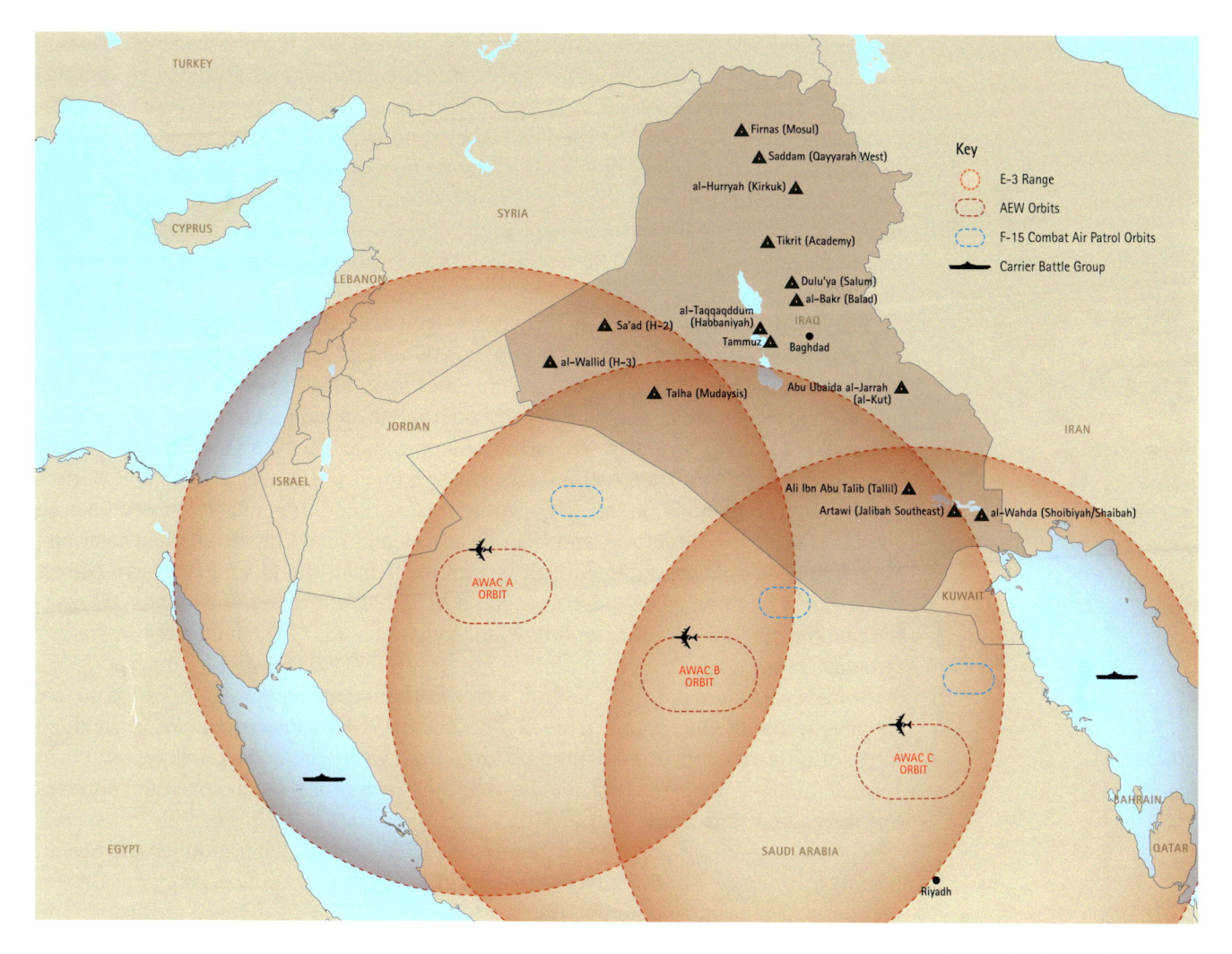

A map showing the E-3 orbits and AOR for Operations Desert Shield and Desert Storm. (James Lawrence)

Horner USAF. Essentially the JFACC had total control of all assets in theatre including US Navy fleet aircraft. This did not sit well with Navy chiefs at USN Central Command (COMUS NAVYAVCENT). They did not cooperate fully with Horner as they saw the fleet aircraft as having the primary role of defending and flying in support of the carrier battle groups (CVBG), and that attacking land targets was a secondary role. Horner was also given a hard time by certain land battle group commanders who, not being aware of Schwarzkopf's planned primary targets, complained that the Iraqi field units in front of their positions were not being allocated sufficient air strikes to erode their fighting strength.

The process of producing the ATO started with the Joint Targeting Coordination Board (JTCB) who would select targets from those predetermined by the JFACC and those nominated by the Army Forces Central Command (ARCENT) and the Marine Corps Forces Central Command (both of whom would have received a target request list to be hit from their lower-echelon commanders on the ground). The JTCB then passed a list of targets on day one of the operation called the Master Attack Plan (MAP) to the planners and targeteers at CENTCOM. They produced an ATO which matched specific targets to units with the correct capabilities to attack the target, and in the process specified which aircraft were to be allocated, how many, what their weapon load-outs would be, their departure times, and their routes and altitudes en route to and egressing from the target. Additionally it allocated which radio frequencies, IFF codes and call-signs would be used and where and when they would rendezvous with tankers for aerial refuelling if it was required. Overall it planned each 24-hour period by the minute for every fixed-wing platform engaged in a mission on that particular day. It did not, however, at that time include specific details of helicopter movements below 152m [500ft] in the AOR, a situation that was to have dire consequences after the war.

The AWACS aircraft were very heavily involved in the ATO as it allowed the mission crews to assess their traffic loadings and who their 'customers' were going to be while they were airborne. The ATO was disseminated to all aviation units in theatre; it was so long and took such a long time to print out that naval units had to have a hard copy of the ATO for the next day flown out to them during the conflict. Once the attack missions had been flown then battle damage assessment (BDA) reconnaissance sorties (that were also planned on the ATO) would be flown and the imagery passed back to CENTCOM for the targeteers and BDA to assess and decide whether the target had been damaged and degraded enough to remove it from the MAP. At any given time there would be three ATOs in existence: the ATO that was being flown, the ATO nearing planning completion for the following day and an ATO in the planning stages for two days hence.

Some air elements of the Coalition were positioned at Incirlik AB in Turkey under the auspicates of Operation Proven Force, whose 7740th Provisional Wing included a trio of E-3As as part of a 110-aircraft force that was dedicated to striking targets in Kurdish areas of northern Iraq, and beyond the reach of forces deployed in countries and aircraft carriers in the Persian Gulf.

The ground operation to eject Iraqi forces from Kuwait, Operation Desert Storm, began on 16 January 1991. It involved a total of 17 NATO and USAF E-3As: 11 at Riyadh and three at Incirlik, assisted by the five E-3s of the RSAF also based at Riyadh. The same three orbits that had been flown during Desert Shield were flown on a constant basis, with a fourth reserve aircraft being airborne to act as an immediate airborne replacement should one of the three primaries fall down due to a lack of serviceability or an inability to refuel. The USAF E-3s operated with an Airborne Command Element, a five-person team aboard in case contact with the Tactical Air Control Centre was lost for some reason. An RSAF E-3 was also airborne in the Riyadh area to co-ordinate a last line of air defence of the capital should it be required. An E-3 was also launched from Incirlik to protect Turkish airspace and to monitor the attack packages operating within the Proven Force AOR. Sentry crews also managed the aerial refuelling tow lines, vectoring damaged and fuel-critical aircraft towards the nearest tanker to allow them to get back to base. AWACS crews also coordinated the initial phases of combat search and rescue (CSAR) should a Coalition aircraft be shot down and its crew eject. Bill Richards recalled a mission during the first night of the war:

'My first mission (one of three I flew in the first 64 hours of the war) was to the western AOR. Several exciting things happened. The first involved the E-2Cs from the Red Sea and two Navy packages that hit the H-2/H-3 complex of airfields. Both were supposed to be controlled by the E-2s with AWACS as backup. The first package generated some MiG activity as they ingressed and two MiG-21s got airborne. The

E-2 controllers did not call them out until I initiated unknown tracks on them (E-2 radar over land really sucked). As soon as they saw the tracks they started calling out the bogies and two [McDonnell Douglas] F/A-18s engaged and shot them both down.

'Tracking friendlies over the target area was tough as they were attacking from all directions and thus egressing in all directions at high speed. This all set the scenario for the next strike package. As they approached the target area they used the same tactic of attacking from different directions at once; they obviously knew that MiGs had been launched against the previous package. As they ingressed to the target there were no MiGs launched but the friendlies once again began to be difficult to track. An F-14 was capped about 20 miles [32km] south of H-2 and he picked up a radar-only track departing H-2 at high speed and he called it out on the radio. The surprising response from the E-2 was 'Sta-Sta-Standby!' I realised this wasn't going well so I took a Mode 4 sweep of the radar-only target that was about eight to 10 miles [12.8-16.1km] south of H-2. A second later the F-14 pilot came back and said in an excited voice, 'Target, nose, 12 miles [19.3km] hot, am I cleared to engage?' and again the E-2 stuttered 'Sta-Sta-Standby' just as my Mode 4 sweep came back as friendly! The F-14's immediate response to the E-2's transmission was 'Fox 1'. I grabbed every frequency I had at my console and screamed 'Cease Fire South of H-2. Friendlies only that area!' I have no idea what occurred after that and hoped that if the F-14 really had fired a Sparrow that he heard me and broke the radar lock. I know several Navy aircraft did not make it back from that attack but have never found out if they were lost to ground fire or fratricide.[5]

'The next event from that first mission was as the second package was egressing [and] they had to hit a tanker on the way out. The E-2 was never able to get them to the tanker and four of the A-6s declared emergency fuel. So I picked them up on 243.0 (Guard) and started vectoring them to Arar airfield. I was able to contact a US air traffic controller in the tower on Guard and was able to brief him and the A-6s on the situation. At that point I split the A-6s up and gave them individual vectors to Arar and set up a nice little sequence. Meanwhile all my traffic on Guard attracted a US F-111 with a hydraulic failure and an RAF Jaguar with a generator out. Both requested vectors to Arar. So now I had six emergency aircraft on guard and on vectors to land at Arar. Thankfully they all made it in safely. I did get a little ribbing the next couple of days for hogging Guard that night and playing ATC.'

Hawkeye in the Storm

Within hours of Iraq's invasion of Kuwait the USS *Independence* and her battle group, which had been exercising in the Indian Ocean, were dispatched towards the Gulf. They arrived in the area five days later, taking up station in the northern portion of the Arabian Sea. The embarked Hawkeye squadron was VAW 113, which immediately started flying sea control sorties protecting the sea-lanes being used by the Coalition sea transports in the vital harbours of Bahrain, Oman, and the United Arab Emirates. The USS *Dwight D. Eisenhower* and her battle group transited the Suez Canal and took up station in the Red Sea on 8 August. Embarked were the Hawkeyes of VAW-121, which flew continuous missions to protect the CVBG from air attack and to enforce

With 93 days in the combat zone, VAW-116 'Sun Kings', exemplified by E-2C BuNo 163027/NE-602, set a new record during Desert Storm. (Mike Freer)

the UN no-trade sanctions against Iraq for the next month. Also in August the CVBGs of the USS *John F. Kennedy* and USS *Saratoga*, with the Hawkeyes of VAW-126 and VAW-125 respectively embarked, began to switch their positions between the Mediterranean Sea and the Red Sea, thereby providing protection to maritime sea lanes between the Gulfs of Aqaba and southern Djibouti, which were being used heavily by commercial vessels and supply ships for the Coalition on their way to Saudi Arabia.

The Hawkeyes of VAW-115 attached to Carrier Air Wing Five aboard USS *Midway* replaced the VAW-113 aircraft in November, helping to provide monitoring of Iraqi and Iranian and vessels in the Gulf of Oman and off the coast of Saudi Arabia. The following month VAW-123 and VAW-124, now respectively based on the USS *America* and the USS *Theodore Roosevelt*, gave two of their Hawkeyes over to form the Carrier Air Wing Eight Embedded 'Detachment Bahrain', based on the island.

The purpose was to provide surveillance radar cover for the northern end of the Arabian Gulf, when the E-3s withdrew to the towlines for AAR as a result; the E-2 missions that were launched three times a day totalled 395 hours of flight time up until 19 January 1991. However, the record for the longest-serving Hawkeye squadron in theatre went to VAW-116, which operated from USS *Ranger*. Flying for 93 days in the combat zone, VAW-116 vectored Carrier Air Wing Two attack sorties during the early hours of Desert Storm, flew several night missions, and completed its tour in March 1991 after flying missions to enforce Iraqi compliance with the terms of the ceasefire.

In total, 29 E-2s were employed in theatre. Between August 1990 and March 1991 they flew a total of 4,790 hours in 1,183 missions during which they guided attack packages into Iraq, controlled an anti-submarine Lockheed S-3B Viking which launched a Mk 82 torpedo, supervised AAR towlines, provided early warning surveillance for their CVBGs, and even picked up unauthorised aircraft trying to fly over the CVBGs that had been chartered by members of the international press.

The war ended on 1 March 1991, by which time E-2s and E-3s were primarily busy attempting to coordinate the hunt for Iraqi 'Scud' surface-to-surface missiles and Iraqi aircraft that were being evacuated to Iran. By the end of the conflict, the US and Saudi Sentries had flown a total of 10,597 hours in the course of 919 sorties that had resulted in claims for 41 enemy aircraft shot down.

Provide Comfort

After the Iraqi withdrawal from Kuwait, the US administration began calling upon Iraqis to rise against Saddam Hussein and remove his government. In reaction to uprisings of the Shi'a in southern, and the Kurds in northern Iraq, Baghdad deployed the remaining units of its Republican Guards Corps and despite ceasefire agreements with US-led allies prohibiting such activity, the IrAF was used to attack insurgents. Despite repeated warnings, the IrAF continued flying attacks and on 20 and 22 March USAF E-3As vectored F-15s to shoot down two IrAF Su-22s and a PC-7 over northern Iraq. By 5 April 1991, the UN Security Council issued Resolution 688, which demanded Iraq end repression of its population, and the following day the US launched Operation Provide Comfort (OPC), aiming to provide relief supplies to thousands of Kurds that were fleeing attacks by the Iraqi military. Controlled from Incirlik AB in Turkey, Operation Northern Watch resulted in imposition of a no-fly zone (NFZ) over Iraq north of the 36th Parallel, which was enforced by US, British and French assets.

A similar Operation Southern Watch was subsequently launched over southern Iraq, south of the 32nd Parallel, in a vain attempt at saving some of local civilian population. Provide Comfort was run by the US European Command (EUCOM), and included 45 aircraft and 1,400 support personnel from the US, UK and Turkey, who provided refugee camps for the Kurds. To provide AEW&C cover for OPC and supporting aircraft the AWACS from the now renamed 552nd Air Control Wing remained in theatre after the end of the war. While Provide Comfort ended in late 1991, the last two operations were to extend for another 10 years, during which several air strikes against targets in central Iraq were launched in response to Iraqi violations of UN resolutions.

The mix of relief and air-policing operations, and a failure in the relevant chain of command, eventually caused one of biggest tragedies to afflict the AEW community. On 14 April 1994, the crew of an E-3A from the 963rd Airborne Air Control Squadron (AACS) operating over northern Iraq detected two Sikorsky UH-60A Black Hawk helicopters of the US Army over northern Iraq. The helicopters were flying from Diyarbakir in Turkey towards Zakho, to pick up a number of passengers including US Army officers and several Kurdish functionaries. On entering the NFZ, the crews of both helicopters had checked in with the AWACS, advising the operators about their presence and their destination. However, the controllers failed to warn the helicopter pilots that they were squawking outdated Mode 1 and 2 IFF codes, or to use a radio frequency common for all allied aircraft in that area. The Sentry tracked the helicopters until they took off from Zakho and flew in the direction of Erbil: then contact was lost because of the mountainous terrain. Meanwhile, two F-15C Eagles of the 53rd Tactical Fighter Squadron USAF arrived in the skies over northern Iraq, but were not advised by the AWACS crew about the presence of US Army helicopters. Around 10.22 local time, an F-15 pilot picked up a radar contact flying slowly between the mountains, some 64km (40 miles) southeast of his position. The E-3 crew did not attempt to identify the contact, and again failed to inform the Eagle pilots of the US Army helicopters in that general area.

After receiving a negative response from IFF transponders on board the Black Hawks, the F-15 pilots decided to make a visual identification pass on the two 'slow-movers'. Despite one of the USAF pilots making a pass only 200m (700ft) above and 305m (1,000ft) away from the UH-60s, he misidentified them as Soviet-built Mil Mi-24 helicopter gunships as operated by the Iraqis. Minutes later, one of the UH-60s was shot down by an AIM-120 AMRAAM, while the other was destroyed by a single AIM-9 Sidewinder. All 26 persons on board the two UH-60s were killed. By 13.15 Kurds on the ground had notified the MCC that they had seen the two helicopters shot down and investigations into the incident were started. A USAF investigation board was duly convened and its findings reported that:

1. The fighter pilots had misidentified the helicopters.
2. The IFF transponders on the F-15s and the UH-60s had not operated correctly for unknown reasons.
3. Confusion existed as to how the OPC air operations procedures and responsibilities applied to MCC helicopters.
4. The AWACS commander was not correctly qualified in accordance with USAF regulations and that the AWACS crew had made mistakes.
5. OPC personnel were not properly trained in the RoEs for the Northern NFZ.
6. The UH-60s were not equipped with radios that allowed them to talk to the F-15s.
7. The shoot down occurred because of a chain of events started by the lack of clear guidance from the CTF command to its component organisations.
8. The AWACS crew did not make any radio calls during the intercept warning the two F-15s of friendly helicopters in the area.

AWACS operators around the world learnt many lessons from this episode, and consequently RoEs and SOPs were tightened in an effort to prevent such a tragedy ever occurring again.[6]

War on Drugs

In November 1981 the US Customs Service (USCS) and the US Treasury launched Operation Thunder in an effort to stop drug trafficking along the east coast of the US. This enterprise was coordinated with the US Navy, which deployed its E-2Cs to track down and help intercept 31 suspect aircraft, resulting in the seizure of drugs worth hundreds of millions of US Dollars. However, despite the benefits of the availability of the Hawkeye's AN/APS-125 radar to find the slow- and low-flying drug-running aircraft that were approaching the Gulf of Mexico from South America, the programme was not continued due to the costs involved and the Navy's assertion that the missions and their profile did not give added value to the training of the Hawkeye crews. On the other hand, the USCS had seen the effectiveness of an AEW capability and was keen to retain this very useful weapon in its arsenal. To that end the USCS executive looked at acquiring such an asset for itself, the obvious initial choice being the Hawkeye. The US Coast Guard eventually managed to obtain two E-2Cs and establish its own Hawkeye squadron, the Coast Guard Airborne Warning Squadron One (CGAW-1), in January 1987.

CGAW-1 was embedded within the Airborne Early Warning Component of the US Atlantic Fleet and nominally based at NAS Norfolk, Virginia, but it regularly operated from a variety of bases further south, as required. Not only did the USCG crews encounter problems with lack of space at Norfolk, the organisation also wanted to become independent from Navy crews that had initially operated the Hawkeyes. After

Chapter 5

sufficient personnel were trained, two additional E-2Cs were acquired in 1989, around the time CGAW-1 moved to St Augustine airfield, Florida.

The Hawkeyes proved ideal for the role, providing a command and control service using their data-link system to relay a real-time picture to the South Florida Interdiction Centre that served as the HQ for these types of operations. The mission was jointly conducted by the USCG, which was responsible for the interdiction of the drug-smugglers' aircraft and marine craft from 12 miles offshore of the country of departure and the US shoreline, and the US Customs Service, which took over from the shoreline to arrival area. The latter could be small landing strips, in which case helicopter-borne personnel would swoop down on these strips and clearings and apprehend the smugglers and seize the narcotics. To assist the USCG Hawkeyes in picking up the slow-moving and surface-hugging targets early they would be forward deployed to airfields in other participating countries stretching from Belize down to Grenada. During their final year of operations the CGAW Hawkeyes were deployed for 293 days out of 365 on anti-narcotics missions.

Sadly a year later on 24 August 1990 the squadron suffered a loss when the crew of E-2C serial number 3501 reported a cabin fire when on recovery to NAS Roosevelt Roads, Puerto Rico, at the end of a sortie. It appears that the crew fought valiantly to try and extinguish the flames but the flying controls were damaged and the aircraft became uncontrollable and crashed, killing all four crewmembers. As a result of this loss and the expense of maintaining the operations of the squadron against declining budgets, the unit was disbanded and the three remaining aircraft returned to the Navy in October 1991.

CGAW-1 E-2Cs were only rarely photographed during the type's four years of service. (USCG)

Prior to being disbanded in 2013, VAW-77's E-2Cs undertook a unique anti-smuggling role.
(Jason Grant)

Carrier Airborne Early Warning Squadron 77 (VAW-77) 'Nightwolves', US Navy.

In October 1995 a new US Navy E-2C squadron, VAW-77, was formed and based at NAS Orleans in Louisiana. The squadron comprised reservist as well as regular personnel and was established with the primary task of suppressing narcotics and people smuggling. For up to six months of the year the aircraft and crews were detached to forward operating bases such as Roosevelt Roads and friendly countries in the Caribbean including Belize and Honduras. These particular Hawkeyes were specially fitted with the Garmin AIS 530 GPS, which was located in a thimble-shaped housing on top of the circular radome. In Fiscal Year 2009, the unit flew more than 2,000 hours and assisted in the seizure of 17.2 tons of cocaine and the rescue of 15 people. VAW-77 was deactivated at Naval Air Station/Joint Reserve Base New Orleans on 9 March 2013.

Blue Sentinel

A long-term solution for the USCG's requirements dictated a platform that had a longer duration and more crew space aboard the aircraft, and resulted in a proposal for a Lockheed P-3 Orion maritime patrol aircraft to be fitted with the Hawkeye's radar installation. The result of a corresponding study was a manufacturer-funded prototype based on conversion of ex-Royal Australian Air Force P-3B A9-299, equipped with the AN/APG-125 radar to serve as sales demonstrator, in 1988. The resulting AEW Orion proved to be a relatively cheap and reliable airframe, a viable alternative to the much more expensive E-3 Sentry. After concluding that the project was feasible, the USCG ordered three examples under the designation Blue Sentinel in early 1988, though they were to be equipped with the improved AN/APS-138 radar.

The initial Blue Sentinel flew in April 1988 and was officially handed over to the USCG at NAS Corpus Christi, Texas the following month. It began flying missions in support of the Air Interdiction Customs Program that had originally commenced in 1971. It was so successful that four more examples were ordered. Despite the acquisition of these five aircraft the US Customs and Border Protection (CBP) constantly had to requisition E-2s from the Navy and E-3s from the USAF to assist them, as those platforms were able to bring the 'control' aspect to the task due to their more advanced radars. The AN/APS-145 radar was eventually retrofitted to all (by now) eight P-3 AEW&C aircraft operated by the CBP which gave them the added advantage of a fully automated environment and vastly improved detection of targets over land as well as sea (a 40 per cent radar range improvement over the APS-138) with a detection range of more than 322km (200 miles). The aircraft has a range of over 7,403km (4,600 miles) and an endurance of around 14 hours. The CBP usually operates the aircraft in the company of a 'slick' P-3A that has been fitted with the AN/APG-67 acquisition radar as used in the F-15 fighter. The AEW&C Orion will locate a potential target at long range as it departs a suspect airfield within its AOR (the Gulf of Mexico, the Caribbean and the international waters off South America), while the 'slick' P-3B Long-Range Tracker (LRT) will usually then be vectored to a position behind the suspect aircraft.

The LRT will then track the target with its AN/APG-67 until it comes into visual range where the suspect can be identified using high-intensity optical devices. Reports are then transmitted back to the CPB control centre via encrypted data-links and the Communications Over-the-Horizon Enforcement Network (COTHEN), which can include real-time long-distance daylight video, electro-optical and infrared imagery. The AEW aircraft operated by the CBP were progressively modified and upgraded throughout the 1990s and into the next decade, as the cost of operating the fleet of eight P-3 AEW&Cs and P-3B LRTs was far outweighed by their ability to locate and prevent people and drugs trafficking in the region.

Late in 1991, the US reinitiated its attempts to deploy military AEW assets to assist other government agencies against drug traffickers. This time E-3As of the 964th ACCS USAF were operated out of a base in Florida and flown in an orbit over the Gulf of Mexico to work with the Drug Enforcement Administration (DEA) Task Force based in Belize in a massive operation that lasted for a week and involved assets from the UK, Colombia, Honduras and Mexico. Ian Shaw was working as the helicopter programmer and tasker in the Tactical Operations Cell (TOC) at Airport Camp Belize at the time:

'A large part of the day-to-day flying operations in Belize at the time involved efforts to assist the Belizean government in their fight against narcotics growing and smuggling in the country that was rampant at the time. The RAF operated a flight of [British Aerospace] Harriers GR.Mk 3s that had been deployed originally in the early 1970s to provide an air deterrent against a potential Guatemalan invasion. During the six months that I was there the Harriers were mainly being used in the mornings as photo-recce aircraft fitted with cameras to overfly known smugglers' airstrips in the jungle areas that were being used as refuelling stop-off points

A night shot of two US CBP P-3AEWs in their original livery at their NAS Jacksonville, Florida home base. (CBP)

A typical Blue Sentinel formation: one P-3AEW and one P-3 LRT, here in their current livery. (CBP)

for twin-props on their way north to Florida with their loads of narcotics. As soon as the Harriers returned, their imagery would be processed by the on-base Reconnaissance Interpretation Centre (RIC) annalists who would look for the telltale signs that strips were intended to be used that night for fuel stops, the indicators being stacks of oil drums, fresh tyre tracks and sometimes large canvas crosses laid out on the strip. If any were found the nearest Belize Defence Force (BDF) unit would be alerted and an ambush set for any landing aircraft that night.

'*Several aircraft were intercepted on the ground and impounded was an attractive collection of stripped-out twin turboprops stored in a wired-off compound in front of the control tower that were periodically auctioned off to interested American buyers a few times a year!*

'*In the exercise in question things became quite intense, with characters that had been seen around the camp in chinos and short-sleeved shirts suddenly appearing in US fatigues and peaked caps – these individuals were actually US DEA operatives in their real jobs. The dark blue armour-plated Piper Thrushes intensified their searches for marijuana fields and the unmarked dark-green [Bell] UH-1 the DEA used became a more regular sight. In the TOC a satellite comms box was installed and several new faces appeared to work with us for the duration of the operation. The Royal Navy West Indies Guard Ship (WIGS) HMS* Amazon *cruised off the Belizean Coast and USCG cutters were strategically placed in the Gulf of Mexico and off the coast of Honduras. All was set for the operation to begin.*

Concurrently in Belize a widely publicised weapons amnesty was announced so those with illegal firearms could hand them in to the authorities 'no questions asked'. The situation came to life during the night times with reports coming through of aircraft getting airborne from a certain South American country with their illegal cargoes being tracked by friendly aircraft. We were told that these latter were USCS

Blue Sentinel P-3AEW N145CS in its new colour scheme as operated by US Customs and Border Protection. (CBP)

Cessna Citations specially equipped with AN/APG-66 radar to shadow the drug-runners as they flew up north across the Caribbean. The whole situation was being controlled from an AWACS orbiting over the Gulf of Mexico whose comms operator kept up a running commentary of what was going on.

'The thing that amazed me the most were the tremendous radio communications problems in Belize. If our [Aérospatiale/Westland] Pumas and Gazelles flew to Ridau Camp around 80 miles [130km] to the south, we had to use HF Morse code as the UHF or VHF did not work at that distance. Yet during the operation the AWACS was hundreds of miles away to the north and when he transmitted it was loud and clear and as if the operator was in the next room. The aircraft that was being tailed that night was very low over the sea but was being easily tracked by the AWACS and followed by the Citation. It appeared that the pilot of the light aircraft realised he was being followed as just off the coast of Belize he appeared to ditch his cargo into the sea and turn around. A cutter was dispatched to the scene to look for the contraband but didn't find it. This whole operation brought it home to me how seriously the US took the situation of narcotics smuggling in the area and especially how good their 'kit' was!'

Balkan Wars

The Balkan Peninsula began a period of turmoil starting in summer 1991, when the former Yugoslavia split into six independent republics. The Serb-dominated government in Belgrade directed military interventions in Slovenia, in 1991, in Croatia, in 1991–1992, and then in Bosnia and Herzegovina, in the 1992–1995 period. While the short campaign in Slovenia resulted in Slovenia's independence from the rump Serbia, conflicts in Croatia and Bosnia and Herzegovina resulted in brutal, merciless and highly destructive wars and ethnic cleansing that were to extend for years.

Early on, the UN reacted by imposing an arms embargo upon Yugoslavia and all the republics that attempted to separate from it. Correspondingly, UN monitors were sent to the region to evaluate and monitor what was going on there. This resulted in involvement of AWACS aircraft to monitor the effectiveness of the embargoes. In 1992,

NATO launched Operation Maritime Monitor with the aim of controlling air and sea traffic in the Adriatic Sea. In order to participate in this operation, E-3Ds from No. 8 Squadron RAF deployed to Trapani air base, on Sicily, while a detachment of Sea King AEW.Mk 2s from 849 NAS was embarked on several RN warships. As more and more AWACS deployed from their home bases of Waddington and Geilenkirchen, Trapani became overcrowded and the RAF component relocated to the USAF base at Aviano AB in northern Italy, a deployment that was to last for several years.

In April 1992, Bosnian Serb forces laid siege to Sarajevo, the capital of Bosnia and Herzegovina, as well as several Muslim-dominated enclaves in the northern and eastern parts of that republic. In October of the same year, the UN passed Resolution 781, imposing an NFZ over Bosnia, and ordering its peacekeeping force UNPROFOR to separate the belligerents on the ground. NATO responded by launching Operation Sky Monitor, re-tasking forces already deployed for Maritime Monitor. They were to enforce the NFZ in order to prevent Bosnian Serb aircraft from bombing civilians. Despite this, flying activity by the Bosnian Serbs continued and in March 1993, the UN passed Resolution 816, this time authorising its member states to introduce 'all necessary measures' to impose a total ban on flights.

Correspondingly, NATO launched Operation Deny Flight. In addition to the NAEWF and RAF E-3s flying eight missions a day, the gloves were now off and NATO fighters could intercept and if necessary shoot down any offenders. Although on the surface this resolution appeared simple, the NATO forces in the region operated under a 'dual key' principle in that before engaging the Serbs they had to obtain permission from the UN Secretary General or his representative in the area at the time, all of which took time and often resulted in no action being taken at all. Furthermore, except for AWACS orbits, NATO now had to fly CAPs and reconnaissance missions over Bosnia, which in turn made necessary the establishment of dedicated tanker towlines.

While Resolution 816 engaged the problem of Serbian flights over Bosnia, it did little to solve the issue of Serbian advances on the ground, especially into UNPROFOR held 'safe areas': the peacekeepers did not have the authority to engage and shoot back, except in self defence.

The E-3 force was now flying eight missions a day to sustain two permanently flown orbits. They used their SATCOM facility to maintain direct links while airborne to the CAOC in Vicenza, Italy and used their data-links to transmit the air picture in real time so that the HQ could keep abreast of what was happening in the airspace over Bosnia. If a suspect flight was detected by the AWACS while on an orbit, the weapons team on the aircraft would vector one of the CAPs on to the contact and a verbal warning would be transmitted on 243.0MHz. The warnings were, as a rule, ignored, as by the time permission to engage was confirmed via the dual-key channels the offender was usually well outside of the area.

Despite the best efforts of the UNPROFOR and NATO, the fighting on the ground did not diminish and in February 1994 measures were taken to try and stop some of the engagements and relieve the pressure on Sarajevo. Weapons exclusion zones were declared around Sarajevo and Goražde, and Serb forces were given an ultimatum to remove their heavy weapons from around Sarajevo by midnight on 20 February, or face air strikes from NATO. Petulantly they complied, however, on 28 February during a Serbian air strike on targets near Novo Travnik two USAF F-16s of the 526th Fighter Squadron enforcing Deny Flight directed and controlled by NATO E-3A intercepted

Chapter 5

Royal Air Force E-3Ds were on hand from early on in the Balkans conflict, initially helping to enforce the arms embargo and control air and sea traffic in the Adriatic. No. 8 Squadron RAF was deployed to Trapani in Sicily.
(Sergio Gava)

and engaged six Serbian SOKO J-21 Jastreb and two J-22 Orao aircraft which were bombing the Muslim Bratstvo arms factory. Warnings were issued to the Serbian aircraft, but were ignored as they bombed the factory and left it on fire. As a result of the blatant ignoring of the warnings and them continuing the attack, the CAOC that was monitoring the engagement immediately authorised the F-16s to engage at 06.45. The first Jastreb was downed and a second pair of 526th FS F-16s that were also being vectored by the NATO AWACS managed to engage and shoot down a further Jastreb, before the Serbian formation reached the border which the NATO aircraft were not permitted to continue across. The remaining Serbian aircraft landed back at their base at Ubdina in the Republic of Serbian Krajina inside Croatian territory.

These were the first combat kills in NATO's history and served to ratchet up the tension with the Serbs. The first successful NATO air-to-ground attack requested by UNPROFOR took place on 10 April 1994, when requests were made for a strike in the vicinity of the Goražde safe area. A Serbian Army command post was hit and destroyed by two USAF F-16s. As a direct result the Serbs took 150 UN personnel hostage on 14 April. Then, on 16 April, the Serbs shot down a Royal Navy BAe Sea Harrier FRS.Mk 1 with a SAM near Goražde; the pilot managed to eject and evade capture. NATO continued ground attacks against Serb Army targets and air bases through to the end of 1994 and into early 1995. On the 2 June 1995 a USAF F-16 piloted by Capt Scott O'Grady, who was one of a pair patrolling over Bosnia, was shot down by a Bosnian Serb SA-6 SAM in a cleverly laid trap. O'Grady safely ejected and managed to evade capture for six days. It is often claimed that O'Grady was located by an F-16 pilot picking up his brief transmissions on his emergency locator beacon radio, but this is not the case. Sources close to the authors have revealed that a NATO E-3A surveillance operator picked up a brief ADF reading from O'Grady around midnight on the night of 7-8 June.

The Nimrod crew also picked up the signal and using their more sophisticated equipment advised the NATO AWACS crew of O'Grady's exact position. The AWACS crew contacted the CAOC in Vicenza, which in turn authenticated the transmissions and confirmed that they were indeed from O'Grady. The command chain was advised that the pilot had been located and a rescue by US Marine Corps heliborne forces coordinated by anther NATO AWACS was set up for 06.35. O'Grady was successfully picked up by a USMC Sikorsky CH-53D Sea Stallion helicopter that flew him out to

USS *Kearsarge* waiting in the Adriatic. Surprisingly it appears that O'Grady has never publicly acknowledged or given credit to the part played by the AWACS crews in his location and extraction.

The war continued unabated and by July 1994 the UN had optimistically set up several 'safe havens', which proved to be far from safe. On 21 June 1995 a Sea King AEW.Mk 2 from 849 NAS was operating over the Adriatic providing organic AEW coverage for NATO ships. Lt Cdr Dave Briggs of 849 NAS recalled that the Sea King monitored an No. 54 Squadron RAF Jaguar GR.Mk 1A operating out of Gioia del Colle depart from controlled flight while engaged in a practice air combat manoeuvring sortie. The American exchange pilot was unable to recover the Jaguar and ejected. AEW Sea Kings have the capability of retaining the original winch system and on this occasion it was used to winch up the Jaguar pilot. The effect of his dripping clothing on the palletised radar gear just inside the door has not been recorded!

Operation Deliberate Force

As a direct result of Bosnian Serb forces overrunning safe areas and massacring thousands of Bosnian Muslims in the Goražde area, the UN and the NATO came under intense pressure to do more in order to prevent any further bloodshed. The next trigger for commencing assertive action came on 28 August 1995, when s Serbian mortar round was fired into the busy market place in Sarajevo, killing 38 and wounding 85 civilians. This time the Bosnian Serbs had crossed the line. The UN 'key holder' at the time was a British senior officer in UNPROFOR, Lt Gen Rupert Smith; he liaised with his namesake, the NATO 'key holder', US Navy ADM Leighton Smith, the commander of Allied Forces Southern Europe. Between them they agreed that after the UN peacekeepers in the vicinity of Bosnian Serb targets had completely withdrawn, NATO would take military action at a much higher and more aggressive tempo.

NATO commenced Operation Deliberate Force, a series of air strikes that were to last until 14 September 1995; strikes took place on 11 days of the period as some days were lost due to poor weather conditions. On the first day of Deliberate Force, AWACS patrolling on the northern and central orbits guided in over 60 strikes. One French Air Force Mirage 2000N was shot down by a Serbian MANPADS 32km (20 miles) southeast of Pale. The two crew ejected safely but were captured by Bosnian Serb forces. Over the 11 days of air strikes, NATO flew 3,515 sorties of all types, involving aircraft stationed as far away the UK and Germany, as well as aircraft from the USS *Theodore Roosevelt* and USS *America* cruising in the Adriatic. Attacks were carried out on 48 different Serbian target complexes and 1,026 bombs were dropped in all.

Sentries played a pivotal role again, this time over much more congested airspace. The E-3 again proved to be versatile and a multi-tasking platform, coordinating the ingress and egress of air strikes, CAPs and AAR towlines as well as CSAR operations. NATO E-3As carried out 74 of the 99 AEW&C missions; the RAF's E-3Ds flew 25 sorties and French E-3Fs the remaining five. They were supported by US Navy Hawkeyes operating from the two aforementioned carriers, which between them flew a further 62 AEW&C missions. Deny Flight was to continue until the completion of peace talks that commenced on 1 November at Dayton, Ohio. By 21 November an agreement had been reached between the warring parties of the Federal Republic of Yugoslavia, Croatia, and Bosnia and Herzegovina. The final 'Dayton Agreement' was signed by all parties on 14 December 1995 in Paris.

On 15 December the UN Security Council dissolved its resolutions that had instigated and supported Deny Flight and the operation was officially terminated on 20 December. Over the 983 days of Deny Flight, 100,420 sorties were flown at the cost of four NATO aircraft shot down. AEW&C activity wound down, but these assets continued to fly sorties in support of the Stabilization Force (SFOR) in the region. The CAOC at Vicenza continued to maintain operational control of AEW assets dedicated to Deny Fight and RAF E-3Ds continued to deploy from Waddington to a small but well organised detachment site at Aviano AB. In the early months of 2007 a financial crisis in Albania bought about a change of government, civil unrest and the deaths of around 2,000 civilians. Once again, the AWACS would be in the vanguard of the operation that followed.

Allied Force

Following the loss of Slovenia, Croatia, Bosnia and Herzegovina, and Macedonia, Yugoslavia was reduced to the state of Serbia and Montenegro. Still dominated by Serbian nationalists, its government subsequently dedicated itself to subjugating the Albanian population of the former autonomous province of Kosovo. Before long, this caused a low-scale insurgency by the Kosovo Liberation Army (KLA). Serbia reacted by launching a 'counterinsurgency' campaign, including widespread reprisals against Kosovo Albanians and, eventually, ethnic cleansing. This time European and NATO leaders were not going to tolerate the Serbian actions, and after the murder of 45 Kosovar Albanians in Racak in January 1995, they acted, demanding that Belgrade permit the deployment of peacekeepers inside Kosovo. Unsurprisingly, Serbia refused and this left NATO with no choice but to use force.

With Brussels lacking the appetite for a protracted ground campaign, the resulting Operation Allied Force was dominated by the deployment of air power; this time Serbia was to be forced to give up control over Kosovo by bombardment alone. Operation Deliberate Force commenced on 24 March 1999, with air strikes against air defence positions and other military bases across Serbia. In an effort to minimise aircraft losses and the handing of propaganda victories to Serbia, strike aircraft were limited to operating above 4,572m (15,000ft), which made accurate bombing very difficult.

During Allied Force, author Ian Shaw worked closely with the RAF E-3D force for the first time. At the time he was asked to go out to Aviano AB, the RAF had been conducting AWACS operations from the Italian base since early 1992, with a forward-deployed detachment of No. 8 Squadron from Waddington. Now, in 1999, they had been joined by crews from the newly formed No. 23 Squadron that was in effect the E-3D Operational Conversion Unit (OCU), but which had put together a number of constituted crews for this campaign. The following narrative focuses on the ground support element that was vital to get the aircraft airborne in order to complete their mission.

'As the situation escalated in Kosovo the staff levels on the detachment at Aviano were also ramped up. The Ops Room at Aviano was situated in a couple of portacabins placed inside a hardened aircraft shelter (HAS) on the north side of the base. In

During the conflicts in the Balkans, Royal Air Force E-3Ds frequently teamed up with Panavia Tornado F.Mk 3 interceptors. (Philip Stevens)

No. 23 Squadron, Royal Air Force.

the cabins was the main Ops Room, a crew briefing room, a tape storage area, and across in the other cabin was the intelligence cell. This setup had been in place for a few years already and operated very smoothly. The Ops Room staff were normally drawn from the Squadron Ops Assistants who were all corporals and were supplemented by Station Operation Room (Stn Ops) corporals as required. I was personally briefed about the Aviano Det the day before I deployed by my good friend Andy Talbot who had just arrived home from Aviano himself.

'*As the situation in Kosovo worsened, the execs at Waddington realised that the Aviano Ops Room would soon need 24-hour manning with both an Ops Officer and a Corporal Assistant. So the bazaars were trawled for those with Ops experience, hence me being deployed on Allied Force. I flew out on a NATO TCA Boeing 707 a few days later. After a rather heavy landing at Aviano a few hours later I now understood why the RAF guys called the TCA 'Take Chance Airways'.*

'*I was on duty that same night from 1700 until 0800 the following day. I was to be dualled up with an Ops Officer who would show me the ropes before I was to go solo the following night. I was teamed up with an experienced No. 23 Squadron Ops Asst, Mark 'Bomber' Harris. Bomber was an expert in proceedings and the requirements of the Sentry crews and the prep they required from us prior to a mission. He soon got me up to speed with how the ATO worked, the flying programme, booking tankers if required with the CAOC, Joint Tactical Information Distribution System (JTIDS) reports, obtaining Notices to Airmen (NOTAMS), weather package compilation, booking diversions, flight plans, diplomatic clearances and how we all fitted into the system at Aviano. The learning curve was steep and the hours very long but I loved it. This was real for the first time in my career this was 'no duff'; even tactical evaluations (TACEVALs) in Germany came nowhere near this.*

'*We all did around a month at a time on the Det except for one of the corporals, Kev Morley, the No. 8 Squadron Ops Assistant. He had been deployed to Aviano on* a four-month Out Of Area posting. This proved a masterstroke as it gave the Det the continuity it needed and if we needed to ask a question Kev was the man. We worked three days 0800 to 1700 followed by three nights 1700 to 0800 followed by three days off (if we were lucky and not called in). The Det Commander was Wg Cdr Di Whittingham, the CO of No. 23 Squadron, a hard but fair man who was intent that the Waddington E-3D wing would put up the best performance possible.

'We had varying numbers of E-3Ds on the detachment at various times. Generally we had three aircraft that would be parked on the 'Strat Pad' on the northern side of the airfield about a third of a mile from our HAS. Sometimes we would be down to two aircraft as the third would be doing a 'frame' swap-over with a replacement aircraft from Waddington. Our serviceability rate was very good; a primary aircraft would be virtually primed and function-checked ready to go at any time, with a back-up spare aircraft. The only major serviceability problems I remember was with the boom air-to-air refuelling [AAR] receptacle that kept getting badly 'dinged' by USAF boom operators in the KC-135s. At one stage Di Whittingham thought there was some kind of conspiracy going on, as for around three AAR fill-ups in a row the boomer damaged our airframes, missing the receptacle and damaging the skin on the cockpit roofs! We did check to see if it was the same boomer that was doing it but it wasn't.

'A mission cycle would start during the early evening when we would get the 'frag' off the ATO, which would designate what missions we were to fly the following day. We called them 'lines'. For instance, if we were fragged to fly two missions then it was a two-line day; occasionally we got a three-line day spread over the 24-hour period. The crews would come in the day before to pre-brief, catch up with the latest Intel and check the special instructions (SPINS) and rules of engagement (RoEs). Generally we were tasked with an early-morning take-off at around 0400 local time. It was our job in Ops to create a mission package for the crew before they arrived at Ops to do the final out brief and then 'walk' to their aircraft. Diversions needed to be booked, the weather and the NOTAMs for those airfields obtained via the internet. We used Google extensively to get the data we needed. We would prepare a weather brief for the orbit and target areas for the crew with a synopsis faxed over from Waddington MET. We would book transport for the crew, collate and distribute authorisation sheets, sort out the passengers for the flight...

'On most missions our aircraft didn't need to refuel as we mostly flew the central or southern orbits. The on-station time was usually about six hours with around an hour's transit to and from the orbit. Quite often though our MAGICs would be tasked by CHARIOT (the CAOC) to extend on station as a follow-on AWACS had gone tech or was late getting off for some reason, and it wasn't unknown for our crews to double-stint on station with a quick top up of AAR in between.

'We had to move quickly if we heard CHARIOT re-task our crew while on station. We would ring the tanker cell at the CAOC and start negotiating for a tanker to top our jet up. We had a SATCOM box on the wall in the Ops Room… We could listen to the packages checking in and out with the AWACS as the missions were flown; some of it was a bit tense at times. The tensest time was on 4 May 1999 when one of our E-3Ds being flown by a constituted No. 23 Squadron crew pretty close in to the action was targeted by a Serbian MiG-29 flown by Lt Col Pavlovic of the 127th Fighter Sqn, 204th Fighter Regiment. Apparently he was unauthorised to fly and took off from Nis without clearance. He kept his MiG down in the weeds initially *before popping up at 12.41 local time. He was picked up and tracked by one of our weapons controllers (WC); she saw him coming straight for the AWACS as a large NATO strike package was egressing the target area. Most had departed but among the last aircraft in the area was a pair USAF F-16Cs. The link manager (LM) managed to contact the egressing F-16s before she handed them over to the WC. She then called the F-16s back in 'hot' as they had been weapons 'cold' on their way to the tanker and were, I believe, pretty short of gas (I later heard the phrase 'they were on fumes'). She successfully vectored them on to the MiG-29 and they engaged it and shot it down at 12.46. There was some discussion between the AWACS crew with one of the tankers on one of the Adriatic towlines whom if I recall correctly 'moved' his track further in towards Kosovo to meet up with the now fuel-critical F-16s. When the E-3D crew landed back at Aviano and entered the Ops Room the relief was palpable. Although not a 'Top Gun' moment the crew were congratulated on an excellent and professional job and immediately phoned the F-16 pilot who I if I recall had also been flying out of Aviano that day. I must admit that episode bought it home to me how vulnerable our crews were and how this wasn't a game or an exercise… the AWACS then and still is a prime target for the opposition to have a go at.*

'A little before that mission we had been alerted as a wing at Aviano to the MANPADS threat in the area. Intel had been received that possible Yugoslav special forces had got on to mainland Italy with some shoulder-launched SAMs and were heading for Aviano. The fast jets changed thecir departure profile and put the burners in and spiral climbed in the overhead to minimise the MANPADS threat. Of course our E-3Ds couldn't do this: they had to climb straight ahead on full power whatever the threat. I remember Wg Cdr Whittingham telling one of the pilots, 'You've got four engines, you'll only lose one!'

'I did two stints as Ops Officer during Allied Force… Some quarters have taken cheap shots at the RAF crews in other publications, reiterating how they had fewer frequencies and fewer radar consoles than the other AWACS involved in Allied Force. I would like to turn that on its head and point out that even if we had fewer consoles and our TDs and WCs had to stay at their screens longer than anyone else, they dealt with their traffic more efficiently and effectively that the other AWACS contingents… I spoke to several different agencies and squadrons while out there and when they found out they were going to be working with an RAF E-3D they were really pleased about it…'

Operation Allied Force lasted for 78 days and ended on 10 June 1999 when the Serbian government finally realised that NATO would not stop bombing and was actually preparing for the deployment of ground forces. During Operation Allied Force eight NATO E-3As with 12 crews from Geilenkirchen, three RAF E-3Ds with four crews from Aviano, five USAF E-3Cs and six crews out of Geilenkirchen, three NATO E-3As and four crews out of Trapani, and finally four French E-3Fs with four crews flying from Avord all provided round-the-clock coverage of the three AWACS orbits. NATO E-3As flew 495 sorties, 59.9 per cent of the total, and spent 2,782 hours on station. RAF E-3Ds flew 161 sorties, 20.2 per cent of the total, and 939 hours on station. USAF E-3Cs flew 106 sorties, 14.2 per cent of the total, and 656 hours on station. The French Air Force E-3Fs flew 49 sorties, 5.7 per cent of the total, and 263 hours on station.

Allied Force threw up some problems for the AWACS fleets. It demonstrated that years of 'watching and listening' while patrolling on UN operations had degraded other important skills such as fighter controlling, AAR controlling, strike package controlling, CSAR and CAS direction. The force had now evolved imperceptibly into air-battlespace managers and their training had to be modified to fit the role. The aircraft themselves needed more frequencies; the NATO fleet needed an ESM capability. Consoles needed to be updated to cope better with the higher workloads and, above all, an operational culture needed to be applied, with training to suit. The problem of having two ATOs – one for US eyes only that contained stealth aircraft mission data, and another one that did not – caused serious flight safety problems in very congested airspace. The story of an F-117A formatting on an E-3A and 'requesting gas' illustrates the point. The E-3s also needed to have a defensive aid suite fitted, as their 'sanctuary areas' were shrinking, with modern potential enemy fighters being armed with longer-ranged missiles. Many of these issues were raised by the squadrons and senior officers made promises, but few 'fixes' came to fruition.

1   Bill Richards, email and website story, May 2013.
2   Brig Gen Ahmad Sadik (IrAF, ret.), interview with Tom Cooper, March 2004.
3   According to Sadik, the IrAF was unconcerned about a possible threat of Iranian F-14s, since the IrAF Intelligence Directorate successfully convinced the air force that these were non-operational since 1981.
4   Because the IrAF still did not operate them, neither of the two aircraft was demanded back from Iran in a famous letter written by the Iraqi Foreign Ministry to the General Secretary of the UN, in September 1991.
5   In a separate interview, some of the US Navy crews involved denied any such incident. Indeed, the CO of the F-14 unit in question stressed that all of his Tomcats returned to their carrier with all of their missiles still where they should have been – on their launch rails.
6   Many valuable lessons were learned from this episode across NATO, some of which were to directly affect author Ian Shaw when he was working as a Helicopter Planner in KFOR 5 HQ in Pristina, in 2001. He regularly had only 82 IFF codes to allocate on a daily basis to 102 helicopters deployed in the zone of responsibility: every single time he had this episode from northern Iraq somewhere in his mind.

Chapter 6

THE WAR ON TERROR, 2001 TO TODAY

As we have seen, AEW&C in the confines of the CONUS traditionally involved keeping watch for Soviet bombers sneaking in over the North Pole, Alaska or the Northern Atlantic. As the bomber threat diminished and the ICBM gained ascendancy, the AEW 'barrier' operations had ceased and the extended DEW Line and Ballistic Missile Early Warning System (BMEWS) Line took over as the sentinels to keep watch for attacks on North America.

Successful USAF AEW&C operations had been conducted over the Gulf of Mexico to direct American interceptors looking for drug-smuggling aircraft and ships coming from South America. AEW&C operations before, during and after Desert Storm had also proved their worth.

But this time the threat came from within US airspace. On the early morning of 11 September 2001, hijackers had taken control of airliners just after take-off from internal east coast US airports and would use them to deadly effect while civilian air traffic controllers and their military air defence counterparts were left confused at what was unfolding before them on their radar screens. Immediately after the first airliners had impacted their targets and it became apparent what was happening, the AEW&C gears went into motion. On that day 19 extremist Islamists motivated and sponsored by the terrorist organisation al-Qaeda had nearly-simultaneously hijacked four passenger aircraft under way over the CONUS, and then used them as 'flying bombs' against selected targets in New York and Washington DC. Two of the airliners were crashed into the World Trade Center, one into the Pentagon, while the fourth failed to reach its target: it hit the ground in Pennsylvania, after passengers attempted to overpower the hijackers.

The first AEW&C operations took place just a few moments after the attacks began. Two E-3s from the 552nd ACW were already airborne that morning on training missions: one off the coast of Florida, another near Washington DC. Following the first hit on the WTC's North Tower, the former was instructed to head toward Sarasota, on Florida's west coast, and follow Air Force One with President George W. Bush aboard, while the Boeing VC-25 flew to Barksdale AFB, Louisiana, then Offutt AFB, Nebraska, and then back to Washington, escorted by two F-16s from the 147th Fighter Wing then stationed at Ellington Field, Texas.

Two major operations, also involving E-3s, followed soon after. The first was a result of the US request to invoke Article 5 of the NATO mutual defence agreement, which committed each signatory to consider an attack on another member of NATO to be an attack on the whole of NATO. As a result, Operation Eagle Assist was initiated, which lasted from 9 October 2001 to 16 May 2002. It involved the deployment of five E-3As from the NATO AEW&C Component HQ at Geilenkirchen AB, Germany.

The five Eagle Assist Sentries (augmented by a further two in January 2002) deployed to Tinker AFB from where they flew sorties planned to enhance NORAD. They meshed into the US system, flying as airborne fighter control posts managing the vastly increased tempo of USAF CAP missions, easing the operational pressure on the USAF AWACS fleet. When Eagle Assist came to an end, NATO's Sentries had logged 4,300 flight hours flown over a total of more than 360 missions.

NORAD was tasked with expanding the number of Air Sovereignty Alert (ASA) missions over US airspace in order to prevent another such attack, in an operation that did not have an end date. Three days after the 9/11 attacks USAF E-3s were put in the air along with fighters and tankers as part of round-the-clock combat air patrols. However, the high costs of this initiative and intense pressure on aircraft and crews soon saw it replaced with ground alert scrambles. In these, the E-3s only took part when required, directing fighters flown by all three US air arms against possible threats. These missions are still ongoing.

AEW&C assets aboard US Navy aircraft carriers were also directly and indirectly involved in operations following the terrorist strikes. The E-2Cs from VAW-125 embarked in the USS *George Washington* (CVN 73) and those of VAW-121 based in the USS *John F. Kennedy* (CV 67) soon left their homeports, along with their respective air groups, to form an advanced barrier against other possible strikes from the Atlantic Ocean. In addition to them, the Hawkeyes flown by VAW-124, part of the air component of USS *Enterprise* (CVAN 65), which was about to return from a six-month deployment in the Persian Gulf, remained on station in the region, being augmented by another four E-2Cs from VAW-117 aboard the USS *Carl Vinson* (CVN 70), which was heading to those waters to join *Enterprise*.

The eight Hawkeyes already in the Persian Gulf during the first days after 9/11 were reinforced by a similar number of E-2Cs equally shared by VAW-112 embarked in the USS *John C. Stennis* (CVN 74) and VAW-123 based in the USS *Theodore Roosevelt* (CVN 71). The US was keen to strike back at the perpetrators of 9/11, and the first to feel their wrath in President George W. Bush's declared 'War on Terror' were al-Qaeda operatives at training bases in Afghanistan, where they received protection from their Taliban cohorts.

US Navy Hawkeyes would be augmented by their French counterparts, from Flottille 4F aboard the *Charles de Gaulle* (R91), which arrived in the theatre of operations as part of the French Operation Héracles in late 2001. Larger AEW&C assets were already stationed in the vicinity of what was to become the combat zone, RAF Sentry AEW.Mk 1 ZH106 *Grumpy* having deployed to Seeb air base in Oman in early September, to take part in a British-Omani joint exercise.

The sole RAF Sentry was reinforced by five USAF 552nd ACW, 960th AACS E-3Cs, which departed from Tinker and headed for Thumrait, also in Oman, on 27 September and started to fly operational sorties two days later. Last but not least, the final Sentries to join the AEW&C build-up for what became known as Operation Enduring Freedom (Operation Veritas for the RAF) were two other additional RAF Sentries from Waddington.

The operational chain

The command and control chain for aerial operations in what was to become Operation Enduring Freedom was centred on the CAOC maintained by the air component of US Central Command (CENTCOM) located at Prince Sultan AB in Saudi Arabia. AEW&C assets were also controlled by the CAOC through the ATO. The 'Bossman', as the Sentries were usually referred to, acted as communications bridges, relaying to the strikers the data needed to perform their missions, although they also received such data from Joint Terminal Air Controllers (JTACs) at CENTCOM. Looking at the chain in more detail, requests for airstrikes were generated by the Air Control Elements (ACEs), attached to the special operations forces (SOF) in the field; the ACEs liaised with the SOF liaison officer located at the CAOC. The next step was to contact the Central Air Force operations directors, who transmitted the order to the air battle managers (ABMs) aboard an orbiting E-3. As the operation progressed, another 'link' was added to that chain: the Air Support Operations Center (ASOC), stationed at Bagram AB, Afghanistan, a formation assigned to the US Army's corps level, with a dual role: advising ground commanders regarding the application of air power and liaison between the US Air Force and Army on the subject. After the ASOC commenced working, the E-3 was placed between the new element and the CAOC.

The French Hawkeyes taking part in the campaign also came under the tasking of the ATO while flying in Afghan airspace. Conversely, the US Navy Hawkeyes received orders from two control authorities, the CAOC at Al Udeid air base in Qatar and the Control and Reporting Centre (CRC), at Kandahar Airfield, Afghanistan.

Enduring Freedom had four phases, the first of which was known as Crescent Wind. This was an aerial campaign planned to neutralise the Taliban air defence systems, command and control centres, airfields and ground-based forces that threatened the US-led led Coalition.

The initial sortie flown by USAF Sentries took place on the night of 7 October, when they controlled 15 B-52Hs and Rockwell B-1B Lancer bombers operated by the 28th Expeditionary Air Wing, which had departed from Diego Garcia, Guam, augmented by two Northrop B-2 Spirits that flew direct from their base in Missouri. Other strike assets included McDonnell Douglas F/A-18C Hornets from the USS *Enterprise* and USS *Carl Vinson* in the North Arabian Sea, which were vectored by the Hawkeyes off their respective ships.

The formation's targets included Taliban ground-based early warning radars and military headquarters buildings and the airfields at Herat, Kabul, and Kandahar, Mazar-e-Sharif and Zaranj. Also targeted were SAM platforms and aircraft. The 'baptism of fire' for RAF Sentries in Afghanistan occurred on the night of 9 October, when they flew missions to vector US strike assets launched to hit the airfields at Herat and Kandahar. With the most important enemy fixed targets destroyed after little more than a week of the strikes commencing, the Coalition turned its attention to the mobile targets. The so-called time-sensitive targets (TST) and dynamic targets (DTs), required observation from assets such as the General Atomics MQ-1 Predator armed UAV, in order to neutralise hostile elements as soon as they were detected, but also the lengthening of the loiter time for the strike aircraft and bombers which now required frequent air-to-air refuelling.

However, these new operational requirements caused an initial difficulty to the air battle managers, as they were not accustomed to simultaneously controlling bombers, fighters, UAVs, tankers, ISR (intelligence, surveillance and reconnaissance) platforms and fire support aircraft, each with its particular flight altitude but frequently operating within the same area as each other, not to mention the constant requests from ground troops for close air support (CAS).

The situation increased in complexity further when perhaps the most complex phase of the Enduring Freedom began. Operation Anaconda lasted from 2-16 March 2002 and planned to seek out and destroy what remained of the Taliban forces. Many of those still operating had managed to escape from previous battles and were scattered throughout the Khowst-Gardez region, near the border with Pakistan. During Anaconda the Coalition AEW&C assets acted in two key ways. First, they dealt with requests for fire support. These were generated by the SOLE, which was the SOF element working in the CAOC, and then passed on to the Sentry, which then directed the orbiting strike aircraft to the target. Alternatively, the Sentry was contacted by the strikers, which were sent to appropriate orbits, where they were then transferred to the responsibility of one of the enlisted JTACs assigned by the aforementioned ASOC. The participating E-3 operational routines were broken up into three eight-hour windows, the USAF E-3Cs taking two, and the E-3D taking the third. The former, operated by the 963rd Expeditionary Airborne Air Control Squadron (EAACS) flew for 15,713 hours distributed in 1,259 sorties between 27 September 2002 and 23 March 2003.

Operation Iraqi Freedom

Operation Iraqi Freedom was initiated by the US on 19 March 2003, with the aim of destabilising, isolating and overthrowing Saddam Hussein in Iraq and providing support to a new, broad-based government in Baghdad. To achieve such an array of aims the US formed another multinational Coalition. In terms of AEW&C assets, when hostilities began these comprised five US Navy E-2C squadrons aboard five carriers and the USAF's 363rd EAACS Sentries, based at Prince Sultan AB, Saudi Arabia (which

A Royal Air Force E-3D from No. 8 Squadron arrives at Prince Sultan Air Base in Saudi Arabia during a sandstorm.
(Ian Green)

also hosted RAF E-3Ds, part of Operation Telic, as the British called their involvement) and the 552nd ACW E-3s, stationed at RAF Akrotiri in Cyprus.

Many of these assets were present in theatre before the hostilities commenced, as they were part of the operations that established no-fly zones in the northern and southern regions of Iraq, Operation Northern Watch and Operation Southern Watch, respectively. These operations helped to greatly diminish the threat generated by Iraqi combat aircraft and SAMs, in such way that only 16 days were required to complete an air campaign that had originally been planned to take 125 days. The overall plan comprised five phases: planning and decision-making, positioning of initial forces, attacking the regime, completing regime destruction, and post-hostilities action.

Command and control hierarchy

As in Enduring Freedom, the CAOC was responsible for the management of the aerial operations along exactly the same lines. This methodology allowed the Coalition Forces Air Component Commander (CFACC) targeting to display incredible flexibility using a combination of the Joint Integrated Prioritized Target List (JIPTL) and TST/DT processes. The TST/DT target types such as 'Scud' missiles, strategic attacks against leadership targets, regime command and control capabilities and Republican Guard contingents, were the most important. Iraqi territory was divided into three zones: North, made up of the cities of Mosul, Irbil, Kirkuk, Samarra and As Sulamaniyah, was targeted by either carrier-based aircraft such as the F-14 or aircraft based in Turkey. The West zone including Ar Rutbah, was to be attacked by Coalition assets departing from different locations, and spearheaded by B-1Bs. The South zone, including Karbala, An Najaf, Al Kut, An Nasiriyah, Al Basrah and Umm Qasr, was to be hit by aircraft departing from Coalition bases, US Navy and Marine assets in the Persian Gulf, as well as some long-range bomber missions from the UK and other countries.

In the course of Iraqi Freedom, allied fighter-bombers flew over 20,000 missions, in which they dropped more than 29,000 weapons, hitting 312 TSTs and 686 DTs. The 363rd EAACS E-3 crews in Iraqi Freedom experienced more dynamic inputs, however, this time the strike packages were much bigger, and the airspace was even more congested than in Enduring Freedom. This was partially due to the greater size of the operation in Iraq as opposed to Afghanistan, as well as the fact that there were three E-3s on station simultaneously over Iraq, where there was only one in Afghanistan. Multiple E-3 operations present unique challenges to crew employment in an already congested airspace. Despite that, from 19 March to 28 May 2003, when the E-3 operations came to an end, the USAF E-3 unit garnered a 100 per cent combat effectiveness rating, totalling 2,198.7 combat and combat support on-station hours, controlling over 24,000 aircraft sorties. During this period of AWACS operations, E-3 ABMs controlled over 700 strike packages and 13,000 individual aerial refuellings, which aided in the destruction of over 150 TSTs.

The four RAF E-3Ds and their nine crews (eventually reduced to six, due to the space constraints at Prince Sultan) reached the same rate of availability, with each crew flying a 12-hour mission every other day to maintain three daily eight-hour on-station sorties in one of the four AWACS tracks over and outside Iraq. Between 12 March and 27 May 2003 they flew 127 sorties (some sources state 112 missions), during which

RAF strike aircraft dropped 679 guided and 124 unguided munitions. The 20 E-2s present in theatre flew 442 sorties, from 19 March to 18 April 2003. On 1 May 2003, George W. Bush declared 'Mission Accomplished', to mark the end of the major operations over Iraq. Coalition involvement in the war in Iraq was to actually last for several more years yet.

Cerberus in action

Over the previous two decades, the Royal Navy Sea King AEW helicopter had matured into a reliable and effective platform that was further improved with a new upgrade that proved its worth in the early days of Enduring Freedom. The Cerberus system was commissioned on 31 July 2002, entering service in February the following year. It was deployed immediately aboard HMS *Ark Royal* as 'A' Flight of 849 NAS and was in position in the Persian Gulf prior to the commencement of hostilities.

The latest Sea King ASaC.Mk 7 was now able to contribute to the formation of a complete and detailed tactical situation of the Iraqi armed forces in the coastal region of Iraq, before the invasion of that country in March 2003. In the first few days of the war the ASaC.Mk 7 played a vital role during the attack on the Al Faw peninsula, coordinating ground attacks by British Army Lynx AH.Mk 7 helicopters and transmitting ground threats to the 3rd Commando Brigade as they pushed into the peninsula intent on capturing port facilities and oil production sites before they could be destroyed by Iraqi Republican Guard units.

Sea King ASaC.Mk 7 XV649 at low level over the sea at North Cliffs at Camborne, Cornwall. (Nick Martin)

The Sea Kings were now capable of locating enemy vehicles on the move, small surface craft and even small inflatable-type boats. The ASaC.Mk 7 was really proving its worth and morale was high, when tragedy struck. In the early morning of 22 March 2003 the two original AEW.Mk 2 airframes were in the midst of a sortie changeover. XV740 with three crewmembers aboard was transiting back to the 'Ark' in good visual meteorological conditions (VMC) at the end of its sortie. XV650 was slightly late getting airborne from the carrier and was transiting out to its appointed orbit station with four crewmembers aboard. For an unexplained reason both helicopters collided head-on and crashed into the Gulf, killing all on board both aircraft.

In December 2006, 'A' and 'B' Flights of 849 NAS were reformed into two new squadrons, 854 and 857, respectively, with 849 NAS now taking over the functions of the training unit for the ASaC.Mk 7. Since then both squadrons have been deployed across the globe on a variety of operations, sometimes embarked and operating from Royal Fleet Auxiliary (RFA) vessels, such as RFA *Fort Austin* during 2007 and 2008 when they were operating in the Indian Ocean searching for and tracking pirates. The two squadrons have also been engaged in the Caribbean tracking drug-traffickers. In 2009 the ASaC.Mk 7s were deployed to Camp Bastion in Helmand Province in Afghanistan, where their ability to track slow-moving vehicles enhanced local commanders' situational awareness. The last two ASaC.Mk 7s were flown back to the UK from Camp Bastion on 17 July 2014.

Operation Active Endeavour

In December 2009, NATO HQ decided to include the alliance's AEW Component in Operation Active Endeavour. This was triggered in 2001 by the 9/11 terrorist attacks and was tasked with securing the Mediterranean Sea and the Strait of Gibraltar (two of the most important zones of maritime commerce) from terrorist interference. In order to fly these missions the NATO FOB at Trapani, Sicily is used by E-3As that have been specially fitted with the Automated Identification System (AIS). This receives a signal emitted by all ships weighing 300 tones or more, while the E-3A's radar transmits in maritime mode and monitors all maritime traffic in that region.

Gathered data, including location, heading and speed of the vessels, is relayed to the Allied Maritime Command Headquarters in Northwood, in the UK. Here, all information gleaned from various sources is fused into a recognised maritime surface picture. If a particular vessel cannot be identified, then an alliance asset, such as a boat, is directed to the scene to take a closer look at the suspect vessel. It is said that more than 2,000 flying hours have been spent on the operation since Active Endeavour began.

Iraq's New Dawn

Between 1 September 2010 and 15 December 2011, 552nd ACW Sentries took part in Operation New Dawn, which covered the withdrawal of US forces from Iraq. During that period they logged 10,000 flight hours and controlled 1,199.2 km^2 (463,000 square miles) of airspace around the clock, providing AEW&C for many assets, including F-16Cs from the 13th and 125th Fighter Squadrons.

A naval flight officer assigned to VAW-121 'Bluetails' tracks aircraft at his station on an E-2C while on a combat mission over Afghanistan.
(Mate 1st Class Jim Hampshire/ US Navy)

Operation Afghan Assist

In late 2010 the NATO Airborne and Early Warning Component (NAEWC) was ordered to deploy to Afghanistan to support the International Security Assistance Force (ISAF) that had been created to establish new security standards and doctrines, and diminish the presence of terrorists in the country. The massive ISAF deployment required the application of many types of aircraft to the task, and the NAEWC was expected to provide air command and control, airspace de-confliction, communications relay and radar coverage in Afghan airspace, aerial refuelling flow management and civil/military aircraft de-confliction.

Initially the deployment involved three E-3As and five crews. The progress of implementing Afghan Assist missions into theatre was divided into three phases. Phase One would comprise a take-off from the FOB at Konya, Turkey, and then a transit towards Afghan airspace. Here, the aircraft would remain on station for around eight hours, after which it would land at Camp Marmal, Mazar-e-Sharif, (the German forces HQ in Afghanistan). The aircraft would return to Turkey the following day. Phase Two would see the alliance's Sentries departing from Mazar-e-Sharif towards Konya, to perform the same on-station operational time. Finally, Phase Three would see all missions start and end in Afghanistan. The first stage was flown during the last two weeks of January 2011, while improved communications and accommodation was under construction at Mazar-e-Sharif. On 30 January the first sortie of Phase Two departed from Camp Marmal, before this stage was finally replaced by Phase Three six months later.

The NATO E-3As currently operate alongside four Royal Netherlands Air Force F-16AMs. The Fighting Falcons are the sole offensive assets operating from Mazar-e-Sharif and perform CAS, with the help of a JTAC, while the E-3A remains on station for up to eight hours. Afghan Assist had completed 10,000 flying hours distributed over 1,000 missions up to November 2013.

Deference and Ellamy

On 26 and 27 February 2011, RAF Sentry aircraft operating from Akrotiri in Cyprus began flying round-the-clock surveillance missions over the central Mediterranean region to monitor Libyan airspace as the situation in that country worsened. The sorties provided cover for UK non-combatant evacuation operations (NEOs) from Libya. Vickers VC10 tankers were also forward deployed to RAF Akrotiri to provide air-to-air refuelling for the E-3Ds, enabling them to remain on station for extended periods. Two UK evacuation missions were also flown into the eastern Libyan desert by RAF C-130 Hercules, which picked up some 300 British and other foreign oil workers and evacuated them.

The E-3Ds continued to operate from Akrotiri to maintain a recognised air picture of Libya in case further evacuation missions were required. Operation Deference was controlled by the UK Joint Force Headquarters, which had deployed into the British High Commission building in the Maltese capital of Valletta on 25 February. Three RAF Boeing Vertol Chinook HC.Mk 2 helicopters were also sent to Malta to augment the growing evacuation force, which now included a ground force element. Two Lockheed Martin C-130Js flew the first evacuation into the Libya desert, carried out without the permission of the Tripoli government, on the afternoon of 26 February. A few hours later an RAF BAe 146 successfully extracted the British ambassador and his staff from Tripoli airport. The two C-130Js and a Lockheed C-130K flew a second non-permissive extraction mission on 27 February. The UK Ministry of Defence revealed that one of its C-130s came under small arms fire during this mission but managed to return safely to Malta.

RAF intelligence, surveillance, target acquisition, and reconnaissance (ISTAR) squadrons, which are normally based at RAF Waddington in Lincolnshire, adroitly changed tack from Operation Deference, dedicated to supporting the evacuation of foreign and UK nationals, to the new Operation Ellamy, which was instigated to support the enforcement of a no-fly zone in Libyan airspace by NATO.

The Sentry provided military commanders with real-time surveillance of Libyan airspace and of the Mediterranean Sea. With its ability to link into a multitude of networks as a communications hub, its capabilities were pivotal to the command and control of RAF and allied combat aircraft that enforced the no-fly zone.

Operation Unified Protector

The situation that in October 2012 culminated in the removal of Col Muammar Gaddafi from the leadership of Libya, which he had occupied since 1969, had its roots in the events of the 'Arab Spring', a popular uprising that began in Tunisia in December 2010 and soon spread throughout North Africa and Middle East. Demonstrations and riots started in Libya in February 2011, when Gaddafi ordered a violent repression against the protestors, during which many died. As a result, the UN passed two resolutions, one establishing an arms embargo on 26 February and another, imposing a no-fly zone, on 17 March, in an attempt to prevent the Libyan Arab Republic Air Force making attacks against the rebels, who had by now titled themselves the National Transitional Council. Six days later the UN Security Council declared Operation Unified Protector to enforce the proposed no-fly zone. Many countries contributed to the campaign with offensive aircraft.

NATO E-3 LX-N90-442, another Unified Protector participant, is being refuelled by a KC-135R from 203rd Air Refuelling Squadron, Hawaii Air National Guard.
(NATO)

As part of the overall mission four other major operations were organised. Operation Mobile was the deployment of Royal Canadian Air Force CF-188 Hornet fighters. Operation Odyssey Dawn was the codename of the use of US assets, and Operation Harmattan was the codename for the deployment of French assets. The British involvement was carried out by the 906th Expeditionary Air Wing based at Gioia del Colle, and the 907th Expeditionary Air Wing operating from RAF Akrotiri on Cyprus, flying three E-3Ds from No. 8 Squadron among other support aircraft.

During Unified Protector, 857 NAS, which was embarked on HMS *Ocean* again, operated in conjunction with Westland WAH-64 Apache AH.Mk 1s to provide them with a live battle-space environment picture over Libya. Four 656 Squadron Army Air Corps Apache helicopters were also present aboard HMS *Ocean* and were supported by a detachment of Sea King ASaC.Mk 7s of 857 NAS from RNAS Culdrose, which completed the UK contingent.

LX-N90459 was one of five NATO E-3As that participated in Operation Unified Protector conducted over Libya in 2011.
(Dr Andreas Zeitler)

A French Air Force E-3F guided the first strikes during Operation Unified Protector.
(Cyrille Cosmao/*Air & Cosmos*)

UK command and control was exercised via the Chief of Joint Operations at Permanent Joint Headquarters (PJHQ) Northwood and Joint Force Air Force Component headquartered at Ramstein AB, Germany. Regarding deployed AEW&C assets, apart from the trio of E-3Ds, airframes ZH101, ZH106 and ZH107, which used the call signs SOLEX 1 to SOLEX 3, also present were three USAF E-3B/Cs, 76-1607, 83-0009 and 80-0138, from the 960th and 964th AACS deployed to Souda Bay, which used the call-sign ZEUS.

French E-3Fs, airframes 202/36-CB *Charognard* and 203/CC *Dogue* temporarily operated from the same base, with the call sign CYRANO. The five NATO E-3As stationed at Trapani were LX-N90442, LX-N90447, LX-N90451, LX-N90458 and LX-N90459, and used the call-sign MAGIC. The French Navy operated two E-2Cs from Flottille 4F, aboard the *Charles de Gaulle* ('2' and '3' under the call-sign DAMOS) and, finally, one EMB-145H, initially '757', from the 380th MASEPE, under the call sign DIGENES, also deployed to Souda Bay.

The coalition was hosted at the Italian bases of Decimomannu, Aviano, Sigonella and Trapani, as well as Souda Bay in Crete. NATO Joint Forces Command (JFC) Naples commanded the operation and air operations were managed from NATO's Air Command HQ for Southern Europe, in Izmir, Turkey. Real-time tactical control was exercised by NATO's CAOC in Poggio Renatico, northern Italy, and was supported by CAOCs in Brummsen, Germany; Izmar, Turkey; Finderup, Denmark and Uedem in Germany. Just prior to the Coalition intervention the offensive portion of the LARAF was comprised of a few old aircraft that were barely airworthy. An additional threat was posed by the Libyan air defence network, whose most powerful weapon was the Russian-made S-200 (SA-5 Gammon), which had a range of 300km (186 miles).

The first strikes

Apart from the dozens of cruise missiles launched from Coalition ships and submarines during the first night of operations on 19 March, the inaugural sorties carried out by the air elements were performed by eight French Air Force Dassault Rafales,

The French Navy provided three elements during the first strikes on Libya: E-2C, Rafale M and Super Étendard.
(Katsuhiko Tokunaga/DACT)

two Mirage 2000-5s, two Mirage 2000Ds, and one E-3F. This formation went deep into Libyan airspace to provide air defence missions to enforce the no-fly zone in the region of Benghazi and to strike military targets identified on the ground that could threaten the civilian population.

The following day saw the combat debut for the Eurofighter Typhoon, when two RAF examples took off from Gioia del Colle to enforce the no-fly zone, with the support of an E-3D. Although most Libyan aircraft were destroyed in the first few hours of the strikes, Rafales destroyed a SOKO G-2 Galeb that was taxiing on the runway, in an engagement supported by one of the French Sentries.

As the war evolved, the AEW&C crews started to control strike packages against Libyan targets, as was the case with the mission flown on 29 March, when a formation of four Rafales, four Mirage 2000Ds and two Super Étendard Modernisés (SEMs), supported by one E-2, carried a strike mission on a military depot located 30km (18.6 miles) south of Tripoli.

On 1 April two AEW&C types, one E-3F and one E-2C, accompanied formations of Rafale and SEMs in missions against targets in Sirte and Misratah. The same support was provided some days later, when the fighters destroyed armoured vehicles operated by the Libyan Army.

Sentries were also instrumental during seek-and-destroy missions to neutralise SAM platforms. Image analysts aboard aircraft including the RAF's Raytheon Sentinel and USAF's Boeing E-8 Joint STARS were tasked with hunting for SAMs. Once a site had been located and confirmed, the information was relayed to the E-3s, where fighter-controllers on board directed the fast jets to destroy the launchers.

On 3 April several Libyan Army armoured vehicles were destroyed around Ras Lanuf by the French contingent made up of joint formations of Mirage 2000D and Rafales, and two joint Mirage F.1CR/Rafale/SEM missions that operated in the area of Ras Lanuf. All these missions were supported by two C-135FRs, one E-3F and one E-2C, which controlled and coordinated the assets.

As the conflict progressed, NATO took direct command of all military operations on 31 March, and most of the American assets withdrew. Kiryakos 'Kirk' Paloulian, a civilian EMB-145 pilot, Hellenic Air Force reservist and researcher provided the following narrative:

Chapter 6

Hellenic Air Force EMB-145H serial number 757 from 380 MASEPE, part of 112 Pterix at Elefsis air base. (Chris Lofting)

380 Airborne Early Warning and Control Squadron (AEW&CS), Hellenic Air Force.

'With the call to arms by the Security Council of the UN (Resolutions 1970/1973) in 2011 to take part in the multinational armada to annex Libya, the US and allies immediately asked the Greek government to allocate fighter aircraft from the Hellenic Air Force [HAF] order of battle in order to enforce no-fly zones inside Libyan airspace. The rate of operations on the Aegean front made such a request impossible. Thus, with this factor in mind, the Greek government elected to participate with a HAF strategic asset of an EMB-145H from 380 Squadron from 112 Wing at Elefsis air base.

'The 380 Squadron detachment stood in readiness in accordance with the NATO Air Tasking Order and the UN Resolution 1973. The Libyan airspace was characterised as the Joint Operation Area (JOA). Under CAOC-5 control, the EMB-145H took part as a command and control asset covering the eastern part of the JOA. In a few instances it was tasked to proceed further to the west in order to provide cover for the north portion of the JOA as the operations and the tactical situation dictated. CC Izmir had given the role of TACON (Tactical Control) to CAOC-5 at Poggio Renatico at Ferrara in northern Italy. Naples was relaying the air traffic radar picture back to CAOC-5.

'During the night of 31 March to 1 April 2011 the HAF Erieye flew for the first time as an AWACS/AEW asset providing the CAOC-5 with aerial coverage of traffic in Libyan airspace, flying in racetrack patterns ranging from areas over the Mediterranean southwest of Crete, to 75 miles [121km] north-northwest of Benghazi. During these missions the HAF AEW operated in orbit areas adjacent to and south of the Athens Flight Information Region (FIR), and within the JOA, as both share a common border to the southwest of Crete. An E-3 AWACS covered the western JOA section.

'In the first phase the 380 Squadron detachment operated in support of the suppression of LARAF fighter activity. The eastern holding areas of the JOA were in the proximity of the operating umbrella of the SA-5 positions; these positions were shown in the latest information briefing cards. There was only one instance of the EMB-145H tracked an active SA-5 site. Other SAMs of lesser range were shown to be active, including SA-2, SA-3, SA-6 and SA-8 batteries, all of which were being targeted by allied SEAD (suppression of enemy air defences) and EW assets. From *the feedback of the Intel reports, recce assets and the usage of ESM, it became evident that SEAD was successful and any fighter activity by the LARAF was minimised.*

'During Unified Protector the HAF detachment took part in support of the following missions: air surveillance, surface picture (SURPIC), recce, DCA (defensive counter-air), OCA (offensive counter-air), SEAD, CAS, ESM and AAR. Specifically, the EMB-145H performed the following: air surveillance/track relay via Link 11/16, SURPIC track relays via Link 11/16. In other instances as ordered by CAOC-5 they directed CAP fighters for visual identification (VID) of Libyan fighters and helicopters. Surveillance of areas for low-speed targets such as Libyan helicopters was also conducted in the areas of Al Jambijia, Benina, Misurata, and Sirte.

'In one incident the EMB-145H was the only AEW asset in the area and there happened to be a slow-mover in the area. Tithe EMB-145H identified it as a Mi-8/17 due to its 'air surveillance/helicopter mode' that can pick up a helicopter rotor even when a helicopter is in the hover. As the tactical director was discriminating the helicopter and the weapons controller/fighter allocator was preparing the friendly fighters to VID and subsequently shoot down the helicopter, CAOC-5 came back with a negative on the intercept.

'Another interesting mission that took place during an afternoon was a chase of a MiG-21 that took off alone heading east, towards the border with Egypt. As there was no data on its intentions, the EMB-145H flew parallel to the MiG, keeping 80nm [148km] lateral distance from it. The MiG-21 was flying at varying altitudes of 2,000-5,000ft [610-1,524m] with no IFF squawking. The HAF fighter allocator directed a pair of French Mirage 2000-5Fs to chase it. The MiG-21 reached what was most probably a southern FOB in the desert where another MiG-23 had diverted and landed.

'They also took operational control of AAR assets when a Turkish KC-135R of 101 Squadron was refuelling four Turkish F-16Cs during day and night-time missions. The EMB-145H also collected ESM data and relayed it via Link 11/16 to other units and also relayed the recognised air and surface picture to the CAOCs.

'In a few instances the EMB-145H showed its value to the Coalition when it was utilised in lieu of the E-3 AWACS being used to transmit geographical coordinates of suspected SAM sites and vehicle convoys.

'The integration of the EMB-145H in Unified Protector was an outstanding success and in due course the commander of CAOC Izmir/Smyrna USAFE Lt General Ralph J. Jodice, an ex-F-15 Eagle pilot, visited the 380 Squadron detachment in Souda to congratulate the HAF aircrew and their mission crew on their performance during the operation. In total the 380 Squadron detachments flew 170 missions, with only one being suspended due to the weather conditions prevailing in the Souda Bay area. The three rotating EMB-145H aircraft that participated in these missions logged some 980 flight hours. The last mission took place on 30 October 2011; the lessons learnt from this operation were most valuable, as was the experience gained working in foreign airspace.'

The last strike sortie in Libya occurred on 20 October 2011, when one of the RAF Sentries orbiting over the no-fly zone detected a large target travelling over the ground at high speed; it called one of the Mirages that were patrolling nearby to investigate. At the same time a US Predator UAV monitoring movements around Sirte spotted a vehicle convoy attempting to flee the city. The convoy was identified as being pro-Gaddafi as it attempted to force its way around the outskirts of the city. Since some of vehicles had mounted weapons, the US drone attacked it with AGM-114 Hellfire missiles. In the first attack, only one vehicle was destroyed, and the others scattered in different directions.

Shortly after the attack, around 20 vehicles of the convoy regrouped again and tried to proceed in a southerly direction. NATO again decided to engage these vehicles. Mirage F.1CRs and Mirage 2000Ds soon arrived on the scene and one of them dropped a 500lb (227kg) GBU-12 laser-guided bomb on the convoy, destroying 11 vehicles. Gaddafi, who had been in one of the convoy vehicles, tried to flee, but he was caught and killed by a crowd of locals.

When Unified Protector came to an end, on 1 November 2011, NATO's E-3As had accumulated 2,125 flight hours distributed between 247 sorties. The E-3Ds logged more than 2,000 flight hours. The French E-3Fs logged more than 2,500 flying hours, including 1,155 in command in the theatre. The two French Hawkeyes flew more than 550 hours distributed among 110 sorties. Finally, USAF Sentries flew a total of 1,650 hours.

Operation Serval

In December 2012, the UN Security Council passed a resolution authorising the use of a military contingent as proposed by the Economic Community of West African States (ECOWAS) in an attempt to expel a number of dissident groups that had invaded Mali nine months previously. The African-led International Support Mission to Mali (AFISMA) had around 3,500 troops available to it; the force received unexpected support from France after the invaders kidnapped seven French citizens in early 2013. The French military launched Operation Serval. The French government declared its objectives as to help the Malian Armed Forces halt and push back the insurgent groups while ensuring the security of civilian populations, and to help Mali recover its territorial integrity and sovereignty. Finally, it aimed to facilitate enough security within the country that two additional international missions could deploy into the state to start their work: the International Support Mission to Mali and a European Union training mission for the Malian Army.

The French deployed two Mirage F.1CRs, six Rafales, six Mirage 2000Ds, and some Aérospatiale Gazelle helicopters. In the first stages of the operations, the fighters were vectored by E-3F 202/CB *Charognard*. The E-3F flew 1,000km (621 miles) from Dakar, Senegal, to Bamako, Mali's capital, from where it carried out its mission.

The operation was broken down into have four phases, of which the halting of rebel forces was from 11 to 15 January 2013, with the first air strike taking place on 11 January by Mirage 2000Ds and Gazelles. The second phase, the conquest, from 15 to 28 January, involved attacks by three Mirage 2000Ds based at Bamako. The third phase, neutralisation, from 28 January to 15 April, saw over 100 strikes against enemy depots and artillery positions. Finally, the security phase took place from 15 April onwards, during which three Rafales and two of the Mirages returned to their home bases. On 24 June, after successfully completing its mission, the Serval forces transferred responsibility for Gao airport back to the Malian Armed Forces. During the operation the Mirage 2000Ds stationed in Bamako flew nearly 450 missions, totalling almost 2,200 flight hours, many of which were performed in conjunction with the E-3F.

Future trends

The distant future may hold the prospect of AEW&C-capable satellites, UAVs or chains of interlinked, advanced fighters with long-range AESA radars. But the major deciding factor in all of these is money. In mid-2014 the Western nations are still emerging from a chain of various financial crashes. As a result, defence budgets have become much squeezed and their priority has been reduced. As this book was being prepared, the US was still in a period of 'sequestration'. As a result, the US military has suffered under large-scale mothballing and early withdrawal of several types, including the disbandment of some US Navy AEW squadrons. The UK is now down to five E-3Ds, with the possibility of further drawdowns. The Royal Navy looks like having a capability gap between the Sea King ASaC.Mk 7's withdrawals in 2016 and the Project Crowsnest Merlins reaching operational capability later in the decade. The NATO E-3A force looks set to be reduced in size in the near future and participating NAEWF nations iare beginning to go their own way. This tendency began with Greece (EMB-145H) and Turkey (Peace Eagle) and may yet include Spain (C295AEW) and Italy (G550 CAEW).

The other side of the coin is those nations seemingly going all-out in an arms race against each other. India, with its A-50EI and EMB-145I programmes, is in competition with Pakistan with its ZDK-03 and Saab 2000 AEW&C. Similarly, Brazil, China, Russia and Singapore are all making efforts to improve air and ground forces, including work on AEW&C capacity. China has made rigorous efforts to develop a whole range of AEW&C assets. While Chinese Y-8Js have been photographed patrolling the disputed East China Sea Air Defense Identification Zone, Japan has responded with an announcement that it will be acquiring additional AEW&C aircraft in the near future.

In the last decades, the AEW&C business has witnessed the appearance of two main, interconnected phenomena: the miniaturisation of systems and the popularisation of electronic scanned arrays. Together, these now permit the development and fitting of compact and effective AEW&C complexes into airframes of all sizes and types. Additionally, there are other ongoing trends in the AEW&C sphere. These are dictated by many factors, such as the increasing use of UAVs for ISTAR roles, uncertainties surrounding the survivability and age of existing AEW&C platforms, and the search for more cost-effective and functional solutions. These trends are distributed into three main areas: the fitting of an AEW&C capability into a UAV; the installation of the early warning portion of the AEW&C complex into a satellite, and the transferring of AEW&C roles to a stealthy, fighter-sized aircraft. But why pay billions to buy stealthy fighter aircraft to act as AEW&C aircraft when the moment they switch their powerful radars into active scan mode they will give away their position to an enemy? And so it seems there is still a future for the traditional manned AEW&C platform, especially as it is still the cheapest option for the role.

Chapter 7

AEW&C OPERATORS TODAY

Australia

Australia began the process of acquiring an AEW&C platform in 1975, when it launched Air Staff Requirement (ASR) No. 64 – Air Defence Early Warning and Control Components that proposed that the Royal Australian Air Force (RAAF) acquire an AEW&C capability, to work in conjunction with other elements of the National Air Defence System including the soon to be introduced very-long-range Jindalee Operational Radar Network (JORN). Five years later, however, ASR 64 was split from the overall requirement and became ASR 77, now solely dedicated to an AEW&C aircraft. The AEW&C and JORN capabilities were always seen as being complementary to each other, with the former enabling the close battle management against the targets disposed over the wide field covered by the latter.

In late 1985 the government took a specific interest in ASR 77, which led to the issue to industry of a request for proposals the following year, after which a request for tenders was issued in late 1988. However, the defence budget was not sufficient to acquire the AEW&C aircraft and ASR 77 was postponed. It appeared again in 1994 as Project 5077 Wedgetail and the Project Definition Study (PDS) which took up the fol-

No. 2 Squadron, Royal Australian Air Force.

The mission compartment in the E-7A Wedgetail.
(Boeing via Marcia Costler)

An E-7 Wedgetail from the RAAF's No. 2 Squadron over the Newcastle and Port Stephens region.
(LAC Craig Barrett)

lowing two years established an extensive list of requirements prescribed to be met by the chosen AEW asset.

The project was officially approved in December 1997; the following month saw the initial design activity contract awarded to three consortiums who had responded to the request for the proposals. They were: Boeing/Northrop Grumman with an adaptation of the Boeing Business Jet (BBJ) equipped with a Multirole Electronically Scanned Array (MESA) radar; Lockheed Martin/Northrop Grumman with a version of the C-130J-30 equipped with an evolution of the AN/APS-145 and, finally, Raytheon/Elta/Airbus, which offered the Elta EL/M-2075 radar mounted on the Airbus A310. The proposal from the consortium led by Boeing was declared the preferred tender in July 1999. In Australian service, the 737 AEW&C is named E-7A Wedgetail.

The Northrop Grumman MESA radar was originally developed in 1988, and incorporated the experience from AWACS in an all-new radar design. It is designed to be lightweight and to minimise impact on the airframe caused by its installation. The antenna is composed of three apertures: two back-to-back dorsal arrays provide 120° of azimuth coverage on each side of the aircraft, and a 'top hat' array provides 60° of azimuth coverage forward and aft of the aircraft. The radar provides air-to-air coverage, air-to-surface coverage, integrated IFF, special track/raid beams, and focused sector operation.

The initial example of the E-7A Wedgetail first flew on 20 May 2004. Originally, it was envisaged that a total of seven Wedgetails would be purchased and delivered from 2005 to 2008. However, the costs of the project meant that the final quantity was reduced to six aircraft. The first E-7A was accepted into RAAF service on 5 May 2010. The last example was delivered on 2 May 2012.

There are two modernisation programs planned to be implemented into the Wedgetail fleet in the future. AIR 5077 Phases 4 and 5A will add an Airborne Mission Segment, Operational Flight Trainer and Operational Mission Simulator and mandatory military/civilian upgrades will be applied to the basic aircraft airframe. Phase 4's initial operational capability (IOC) is yet to be defined, while Phase 5A IOC is scheduled to occur between Fiscal Year (FY) 2017–18 and FY 2019–20. Serving with No. 2 Squadron at RAAF Base Williamtown, the E-7A Wedgetail is expected to have a service life of more than 25 years.

Brazil

In June 1994 the Brazilian government began a process to procure an AEW&C platform and an environmental monitoring aircraft, and cost savings dictated that both tasks be conducted by the same aircraft. A paper was issued by the Brazilian Air Force (Força Aérea Brasileira, FAB) in December of that year that suggested the modification of a Boeing 727 to be equipped with the Swedish active electronically scanned radar, the Ericsson PS-890 Erieye. The system had flown just months before. The following year, the Embraer ERJ-145LR regional airliner was chosen as the winning aircraft for the AEW&C airborne radar programme and in 1998 five sets of PS-890 radars were ordered to equip the aircraft.

During the conversion of the ERJ-145LR into the EMB-145SA (Surveillance Aircraft), the fuselage and wing were reinforced in order to support mountings for the long dorsal radar antenna in an effort to minimise its movement. Two large ventral strakes were also installed under the rear fuselage to increase stability and combat yaw caused by the radar assembly. The antenna was installed at an angle to allow the radar to remain aligned with the horizon as the aircraft flies at a high angle of attack to reduce airspeed and extend sortie duration. After flight tests winglets were added, together with upper and lower vertical finlets on the horizontal stabilizers; the finlets also house the data-link system antennas.

A mission crew consists of a tactical coordinator, electronic warfare operator and an intercept controller.

The Erieye radar has been modified to meet the needs of the new operating environment: an extended early warning capacity was incorporated, by means of a reduced scanning speed, as well as lowering the minimum Doppler speed detection limit, which allows detection of slow-moving contacts such as boats. The scanning width of the antenna is 150° on each side of the aircraft.

The first flight of the EMB-145SA prototype took place on 22 May 1999. Delays in the delivery of the mission equipment and its integration work resulted in the first operational airframes – which received the FAB service designation R-99A – not officially being delivered until 24 July 2002, to the FAB's 2°/6° Grupo de Aviação, 'Esquadrão Guardião' (2nd Squadron of the 6th Air Group, the 'Guardian Squadron') at Anápolis. The remaining airframes were delivered in December of the following year. Although based in Anápolis, the E-99s (as the R-99As were renamed in June 2008) is sent to any part of Brazilian territory where there is a report of illicit air traffic. Typically, two Embraer A-29 Super Tucano attack aircraft, from one of the three squadrons of the 3° Grupo de Aviação, take-off and conduct an in-flight control and warning mission, being vectored by the E-99, whose interception controller contacts the military operations centre at the closest air base and try to identify the approved flight plans of the contacts detected by the E-99. Once confirmed as 'unknown', the detected aircraft becomes the target of the usual interception procedures, which include the interception by the Super Tucanos to visually identify the aircraft and its registration. This may result in the shooting down of the illegal aircraft, as approved in Brazilian law. Hundreds of such sorties have taken place, which has forced the drug barons to forgo smuggling by aircraft for trucks and cars as a means of transporting narcotics.

The E-99 has also been used in the monitoring of air traffic during large events hosted in Brazil, as was the case with the Confederations Cup in June 2013, when the

2°/6° Grupo de Aviação, 'Esquadrão Guardião', Brazilian Air Force.

Brazilian Air Force E-99 serial number 6702 touches down at Natal in 2008 during the CRUZEX execise. (Chris Lofting)

Brazilian Air Force E-99 (EMB-145 AEW&C) serial number FAB6700 flying over the Amazon.
(Embraer)

E-99 managed the airspace around the cities where football matches were played. The airspace was divided into exclusion zones, with the E-99 vectoring Dassault Mirage 2000Cs and Northrop F-5EM fighters, A/T-29s and Sikorsky S-70 helicopters. At least two aircraft were intercepted during the event.

A contract has been signed with Saab to improve the radar antenna and its search and track capabilities and also to develop a new mission compartment. The new variant will have five, lighter and more compact workstations instead of the current three, one of which will be exclusively dedicated to air defence missions. All will be provided with flat-screen monitors, probably arranged in the same configuration as the EMB-145I's mission compartment. Atech will provide the software to enhance the air defence capability; Elbit will be responsible for a new self-defence suite as well as for the new COMINT/SIGINT set, which will offer better accuracy on the location of emitting sources than the device currently in use. Macron is tasked with the ongoing development of the Link BR-2 data-link, Rohde & Schwarz will provide the R&S MR6000R Software Defined Radios which will use the aforementioned Link BR-2 and, finally, Thales will install TSC 2050 IFF transponders.

Chile

Chile currently has a single AEW&C aircraft that was designed and fitted out by Israel Aircraft Industries utilising an ex-Lan Chile Boeing 707-320C which was flown to Israel in 1991 for conversion. It was to be the first aircraft to be fitted with the Phalcon AEW system and was seen after its arrival wearing the Israeli civilian registration 4X-JYI. It made its first flight after conversion on 12 May 1993. Flight testing continued for the next year after which the aircraft was returned to Chile. In Chilean Air Force (Fuerza Aérea de Chile, FACh) service it is named Condór. The operating unit is Grupo de Aviación No. 10 at Aeropuerto Internacional Comodoro Arturo Merino Benítez, near Santiago.

Chapter 7

Chile's sole Condór, seen here over the Andes, is set to retire in 2015.
(Cees-Jan van der Ende)

The Condór does not have 360° coverage from its AESA radar as there are no rear-facing antenna. It has a large bulbous nose radome that houses a 3m (9ft) wide antenna and two cheek-mounted flat-panel antenna on the forward fuselage. This antenna array gives the Condór around 260° coverage orientated to the front of the aircraft. There were initial complaints made to IAI by the FACh regarding the system's poor performance but after remedial work by the contractors these issues appear to have been resolved.

Inside the fuselage there are 11 side-facing consoles for the various operators, 1 and 2 are used to monitor and test the mission equipment, 3 and 4 are for the management of the operation of the radar but have no CRT displays. Positions 5 and 6 manage the ESM/ELINT systems and 7 is the tactical coordinator's position and 8, 9 and 10 are equipped with high-resolution screens and are the actual radar operator's screens. Positions 11, 12 and 13 are the communications management and data-link controller's consoles.

Recent upgrades have seen the addition of the EL/L-8382 ELINT system (replacing the EL/L-8300), as used aboard the Wedgetail, while the TF33-100 turbofans have been replaced with JT8D models. However, due to the system's high running costs the Chilean government has planned to withdraw the aircraft from service in 2015.

Grupo de Aviación No. 10, Chilean Air Force.

China

In September 1969, the Chinese government established Project 926, which was intended to create a viable system for controlling jet fighters over the mountainous southeastern coast of the country where there were no GCI radar stations. It was hoped that the new system would provide radar coverage equal to 40 ground stations and detect up to 24 targets simultaneously with have a radar range of around 250km (155 miles). A People's Liberation Army Air Force (PLAAF) Tupolev Tu-4 bomber, 'Red 4114', was selected and after the removal of its bomb sights and turrets it was converted to house the new systems and equipment derived from the project.

147

Equipment included VHF/UHF radios, data transmission equipment, A-type and P-type displays, and a rotodome that was 7m (22ft) in diameter and 1.2m (3.9ft) thick, which was mounted above the aircraft's fuselage in order to house the Type 843 radar system. The completed aircraft was designated Kong Jing 1 (KJ-1) and commenced a series of flight evaluations in June 1971. According to the most reliable sources an actual working radar was actually never installed in the KJ-1, since the radar and its equipment proved far too heavy. Furthermore the radar itself failed to meet expectations and the KJ-1 did not meet PLAAF requirements, even when fitted with more powerful turboprop engines. The programme was cancelled before the end of 1972, the airframe being retired to a Beijing museum in 1991.

Although an AEW version of the Shanghai Y-10 airliner was considered in the meantime, some 20 years would elapse before the PLAAF resumed its programme to develop an indigenous AEW&C system in earnest. In 1992, a high-level Chinese commission toured aviation manufacturing facilities in Israel, Russia and the UK, looking for an 'off the shelf' system to meet their AEW&C needs. With neither Russian or British systems readily available, an Israeli solution based on the EL/M-2075 Phalcon system was considered a good starting point for developing an indigenous domestic product.

China once planned to develop an AEW&C aircraft on the basis of the Il-76 but employing a configuration similar to that of the Chilean Condór, with nose and tail radomes as found on the Nimrod AEW. However, a more conventional rotodome-based solution designed to give complete 360° radar coverage was chosen instead.

The Israeli connection

In May 1997 a contract was signed between Israel and China to provide four examples of the new aircraft that was to be designated A-50I. In comparison to the standard Beriev A-50, the Beriev A-50I prototype had a fixed 11m (33ft) rotodome that housed the three static AESA antennas, each of which covered a 120° arc, mounted on pylons above the fuselage behind the wing trailing edges. The horizontal strakes fitted above the main gear sponsons on the standard A-50 were removed and replaced by canted rectangular strakes under the extreme rear fuselage; the air intake at the base of the leading edge of the vertical fin was also deleted. A SATCOM fairing was installed on the top of the forward fuselage and internally the modifications included the installation of 10 workstations for the systems operators behind which was a rest area with nine seats, a galley and a lavatory.

After the airframe modifications had been carried out by Beriev, serial number RA-78740 flew from Russia to Israel on 25 October 1999 for the avionics and radar systems conversion and installation work to be carried out by Israel Aircraft Industries' Elta electronics division. Interestingly, RA-78740 was a former Russian Air Force A-50 (formerly 'Red 44', c/n 0093486579), therefore featured a nose-mounted aerial refuelling probe. Unfortunately, China's planned acquisition of a modern AEW&C platform was too much for the US to tolerate as they believed that the system would alter the balance of power between Taiwan and China in the Taiwan Straits. After massive pressure was applied to Israel which included the threat of sanctions from the Clinton administration, Israel pulled out of the contract with China on 12 July 2000. After the removal of all the radar and avionics in Israel the prototype was flown back to China in 2002, where it was utilised in an indigenous Chinese programme for an AEW&C system – reportedly designated Project 998 – led by No. 603 Institute, Xi'an Aircraft

The KJ-1 on museum display. During the 1970s the original Shvetsov ASh-73 radial engines reached the end of their service life and were replaced by more powerful and reliable Zhuzhou WoJiang-6 turboprops. (FYJS-Forum)

The KJ-2000 fleet is operated by the PLAAF's 76th Electronic Warfare Regiment, part of the 26th Special Mission Division based at Wuxi-Shuofang near the Taiwan Strait and facing Japan and Taiwan.
(CDF, via Andreas Rupprecht)

Industrial Corporation (XAC) and the 14th Institute that became the KJ-2000 (Kong Jing 2000).

Its new modifications included the installation later that year of a radar designed and developed by the Nanjing Research Institute of Electronic Technology/14th Institute, presumably similar to the Israeli Phalcon system. The revised aircraft flew for the first time on 11 November 2003. It was followed by four others, all of which lacked the prototype's refuelling probe since they were all Il-76MDs that had been formerly operated by China United Airlines. It is suggested that the Chinese technicians working on the KJ-2000 project received design expertise on the T/R modules, rotodome composite materials and radar and databus from Israel, even after the cancellation of the contract. The KJ-2000 was officially inducted into the PLAAF inventory in late 2005 and most likely reached full operational status in 2007. Given the NATO reporting name Mainring, it flew its first internationally published missions during the 2008 Beijing Olympics.

KJ-200 'Balance Beam'
In the late 1990s China launched a new project that established the criteria for an AEW&C version of the Xi'an Y-8 transport, the Gaoxin 5 (New High 5). The actual project designation remains unknown, but the resulting aircraft is commonly known as the 'Balance Beam'. The project became necessary since the contract to acquire additional Il-76MDs for AEW conversion fell through, meaning a substitute was required for the KJ-2000. The first prototype was a test bed for the Y-8 'Balance Beam' and was based on a converted Y-8F-200 transport aircraft (serial number B-57GL), that performed its maiden flight on 8 November 2001. After an extensive redesign the first true KJ-200 was rolled out in late 2004 and completed a successful maiden flight on 14 January 2005, wearing yellow primer and the serial number T0673. In contrast to the test bed it was based on the improved Y-8F-600 airframe, which features a redesigned

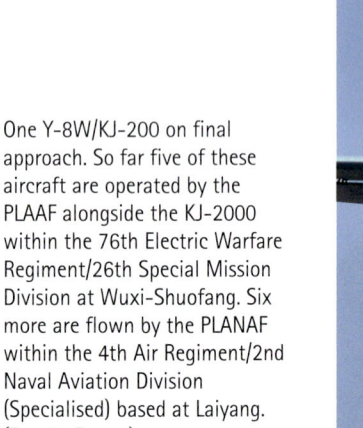

One Y-8W/KJ-200 on final approach. So far five of these aircraft are operated by the PLAAF alongside the KJ-2000 within the 76th Electric Warfare Regiment/26th Special Mission Division at Wuxi-Shuofang. Six more are flown by the PLANAF within the 4th Air Regiment/2nd Naval Aviation Division (Specialised) based at Laiyang. (Top.81-Forum)

Y-8J serial number 9301. (Top.81-Forum)

fuselage with a solid nose and a new tail section with the loading ramp removed. T0673 was equipped with electronic antennas on the nose, tail and wingtips. Its AESA radar was pylon mounted on an 'Erieye style' beam structure above the fuselage and like the Erieye was canted down at the front to adjust for the angle of attack of the aircraft in flight. Two prototypes were initially evaluated by the PLAAF, and reports suggest that on 3 June 2006, T0763 crashed in adverse weather conditions. As a result follow-on aircraft had their fuselages strengthened and lateral stability improved by the addition of vertical fins on the horizontal stabilisers.

The project resumed in 2007 after additional refinements and the aircraft finally was inducted into service. The AESA radar was developed by the 38th Institute and some sources assume some Israeli industry cooperation and assistance. At the time of writing, 11 KJ-200s were operational: five with the PLAAF and six with the People's Liberation Army Naval Air Force (PLANAF). The KJ-200 is also bears the alternative designation Y-8W, while in People's Liberation Army Navy (PLAN) service it is designated as the KJ-200H or Y-8WH.

Y-8 AEW

A further version of the Y-8 appears to have been developed as a back up to the KJ-200. Indeed, it is likely the very first rotodome Y-8 was considered prior to the A-50I and later evolved into a project that ran in parallel and in competition to the KJ-200. The Y-8 AEW prototype had a simpler rotodome configuration and was based on the earlier Y-8F-200 airframe (Y-8CE serial number T0518). The radar comprised a twin array in a rotating radome. After the KJ-200 was selected for production, this aircraft was later used as a factory test bed and in 2013 received a new radar similar to that of the KJ-2000 with three arrays. This then evolved into the KJ-500 described below. After the rotodome-equipped Y-8 AEW became surplus to Chinese requirements its development was continued for the export market. It was sold to Pakistan in 2009 under the designation ZDK-03. The Pakistan Air Force ordered four examples of the type, of which three had been delivered as of September 2014.

Chapter 7

This series-production – or at least pre-series – Z-18J was noted on deck of the aircraft carrier *Liaoning* in February 2014.
(FYJS Forum)

KJ-500

In late 2013, images of a new AEW&C aircraft based on the Shaanxi Y-9, a stretched version of the Y-8 transport, started to appear on the Chinese internet. Designated KJ-500, it is described as a next-generation medium-sized AEW&C prototype, perhaps also designated Gaoxin 10. The radar is reportedly a system developed by the 38th Institute (as with the KJ-200) and during its development several rotodomes of different shapes were evaluated on board a Y-8CE radar test bed. In addition to these three arrays it features large nose and tail radomes that could house additional radar antennas to cover both forward and rear hemispheres. Apparently related to the KJ-500 programme, the original Y-8 AEW rotodome test bed/ZDK-03 prototype has been fitted with a new fixed radar array similar to that on the KJ-2000. The new antenna appears notably wider than that of the ZDK-03. The KJ-500 will reportedly enter Chinese service and as of summer 2014 two prototypes were under test at Xi'an-Yanliang.

One of the two KJ-500 prototypes noted to date.
(CDF via Andreas Rupprecht)

Chinese naval AEW&C

Prior to PLAAF interest in AEW, in 1996 the People's Liberation Army Navy (PLAN) established Project 515 to introduce an aircraft equipped with radar to be used in coastal missions against illegal maritime traffic. The British government sold to China around eight sets of the Racal Skymaster radar, an evolution of Searchwater used aboard the Nimrod MR.Mk 2 and Sea King AEW.Mk 2. Under the designation Y-8J, the radar sets were adapted and mounted inside huge bulbous chin radomes on modified Y-8Cs that had the rear gunner position and loading ramps removed. The main cabin was pressurised, an 8kW generator installed for powering the new equipment and two workstations were added, one for detection and identification of contacts and the second for a fighter controller for managing interceptions. The prototype of the radar-equipped aircraft flew for the first time in September 1998 and three years later, during naval manoeuvres was observed by other nations, it became clear that the Y-8J could act in concert with the Harbin Z-9 helicopters to detect and attack surface vessels operating beyond the horizon via a data-link between the two aircraft.

The JZY-01 is thought to be a structural and aerodynamic test bed for follow-on designs for a dedicated shipborne AEW&C aircraft under development at the No. 603 Institute/XAC. In April 2005 a model of such a type was partially unveiled during a visit to the No. 603 Institute by China's Vice Prime Minister.
(CDF)

One of nine Ka-31s in PLANAF service.
(Top.81 Forum)

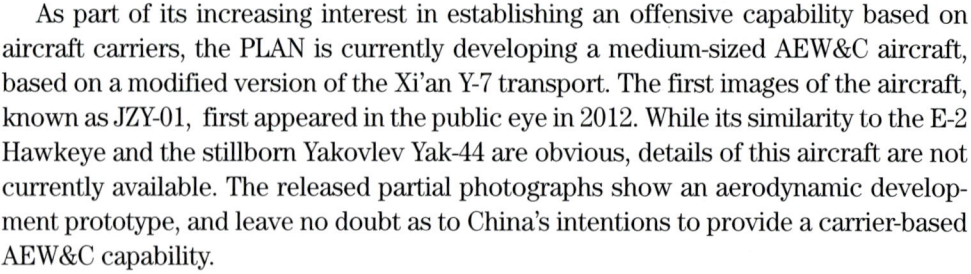

As part of its increasing interest in establishing an offensive capability based on aircraft carriers, the PLAN is currently developing a medium-sized AEW&C aircraft, based on a modified version of the Xi'an Y-7 transport. The first images of the aircraft, known as JZY-01, first appeared in the public eye in 2012. While its similarity to the E-2 Hawkeye and the stillborn Yakovlev Yak-44 are obvious, details of this aircraft are not currently available. The released partial photographs show an aerodynamic development prototype, and leave no doubt as to China's intentions to provide a carrier-based AEW&C capability.

China has also developed an AEW&C version of the modernised Changhe Z-8F helicopter equipped with a retractable radar antenna attached to the rear-loading ramp. The radar is a long-range multi-mode AESA type developed by the 38th Institute. Receiving the service designation Z-18J Black Bat, the aircraft may be intended as a stopgap pending availability of the JZY-01, or may be a rival to the fixed-wing type.

In the meantime the PLAN ordered nine Ka-31s in 2010, which were delivered in 2011 to become the service's first dedicated shipborne AEW asset. The Ka-31s are also capable of operating from smaller naval assets such as the Luyang I/II and Sovremenny class destroyers. The Ka-31 is considered a low-cost stopgap measure until the indigenous Z-18J and dedicated fixed-wing AEW&C aircraft enter service.

Egypt

Egypt purchased E-2C Group 0 Hawkeyes in the mid-1980s, and the initial five aircraft entered service with the Egyptian Air Force at Cairo West in 1987. A single E-2C Group II was purchased in 1991 for evaluation purposes. In 2008, the Egyptian Air Force signed a contract for updates to be incorporated into its Hawkeyes to bring them up to Hawkeye 2000 standard including AN/APS-145 radar. A total of seven Hawkeyes are now operated by 87 Squadron, part of 601 AEW Brigade at Cairo West.

A rarely seen view of an Egyptian Air Force E-2C taken from a video.
(via Pit Weinert)

France

France became the final foreign customer for E-3, and the Armée de l'Air was given a demonstration of the aircraft and its systems in June 1982. Nevertheless, it was not until February 1987 that France went ahead and signed a corresponding order for three E-3F Sentries, while a fourth aircraft was added to the order two years later. The

E-3F serial number 204/702-CD with markings to celebrate 50,000 hours of service by the Armée de l'Air AWACS fleet.
(Chris Lofting)

One of Flottille 4F's E-2Cs lines up for landing on the aircraft carrier *Charles de Gaulle*. (Katsuhiko Tokunaga/D.A.C.T.)

first E-3Fs therefore did not France until October 1990, and officially entered service with the Escadron de Detection et de Commandement (ECDA) 36 'Berry' in June 1992.

In June 2013 Boeing announced that its subcontractor Air France Industries (AFI) had commenced an upgrade of the mission hardware of the first of the four E-3Fs. The mid-life upgrade contract was worth USD324 million and is modelled on the Block 40-45 upgrade that is being applied to the USAF's AWACS fleet. The modifications increase the fleet's surveillance, communications and battle management capabilities. Work on the first aircraft was completed in February 2014 and it completed testing at Avord in 2014. Boeing delivered the first upgraded E-3F in July 2014. Modifications to the other three E-3Fs are scheduled for completion by the third quarter of 2016.

French Hawkeyes

The French Navy took delivery of its first E-2C Hawkeye in 2000, and the three aircraft are operated by Flottille 4F, based in Lorient. Since 2001 exercises have regularly taken place with the US Navy which involve 'cross decking' between US carriers and the French Navy's *Charles de Gaulle*. Since delivery, the three French Hawkeyes have been upgraded to maintain their commonality with their US Navy counterparts. In a USD34.5-million US Navy contract, Northrop Grumman Corporation will modify the French Hawkeyes with an upgraded IFF system, further increasing commonality and interoperability with US Navy E-2D aircraft. Included in the upgrade was the installation of AN/APX-122A IFF Mode 5/Mode S interrogators and AN/APX-123 IFF Mode 5/Mode S transponders. The French E-2s are now at Hawkeye 2000 standard and are equipped with the new carbon fibre eight-bladed propellers.

Flottille 4F, French Navy.

Escadron de Détection et Contrôle Aéroportés (EDCA) 36 'Berry', French Air Force.

Greece

The acquisition of an AEW&C aircraft by the Hellenic Air Force (HAF) was a long and convoluted process. In 1998 the HAF called for tenders for a specific AEW&C platform. Proposals were received from Lockheed Martin (C-130AEW), Northrop Grumman (E-2C Hawkeye 2000) and Embraer (EMB-145SA). In December that year, the Brazilian candidate was declared the winner of the competition and a contract between

Hellenic Air Force EMB-145H serial 729/SX-BKQ at Tanagra air base in 2008.
(Chris Lofting)

A view of the mission compartment in a Hellenic Air Force Saab 340 AEW&C.
(Hellenic Air Force)

the Greek government, Embraer, Ericsson and Thales was signed for the construction and delivery of four AEW&C aircraft.

The standard EMB-145SA was to undergo a comprehensive modification program before delivery including an upgraded version of the Erieye radar that was capable of tracking up to 1,000 contacts in a 360° scan. The aircraft also received ESM/SIGINT capability. The Greek aircraft is known as the EMB-145H (for 'Hellenic') and has a data-link system that uses the standard NATO Link 16. The HAF requirements resulted in an internal configuration similar to the E-99: the rear fuselage has four fuel cells (instead of the six carried by the E-99), a toilet, APU, energy generator and electronic equipment racks. The centre section accommodates five workstations (similar to the E-99's) positioned on the starboard side; they include the SIGINT specialist, radar operator, mission commander and two weapon controller positions. Between these work stations and the cockpit there are two rest seats and a galley.

The contract was signed in 1 June 1999; it included the provision for the HAF to lease two Swedish Air Force Saab S100B Argus aircraft. These were leased from late June 2001 until soon after the Olympic Games of 2004 in Athens.

On 29 October 2004, EMB-145H c/n 671 was delivered by Embraer to the Ericsson Microwaves plant, where the electronic systems were integrated, in a process repeated for each new EMB-145H. Since then four EMB-145Hs have been delivered.

Serial number 304 was one of two Saab 340 AEW&Cs leased by Greece from Sweden between 2001 and 2004.
(Anthony Mylonakis)

Chapter 7

India

The Indian Air Force (IAF) began looking for an AEW&C capability in the 1970s and by the mid-1980s it had started Project Guardian, later renamed Airawat, also known as Airborne Surveillance Platform (ASP). The proposed solution was a radar in a rotodome capable of detecting around 50 targets simultaneously, to be installed on Hawker Siddeley HS748s serving with the IAF. Project Guardian was run by India's Defence Research and Development Organisation, and an HS748 AEW testbed was operated by the DRDO's Centre for Airborne Systems. However, as the programme advanced, it faced some difficulties and resulted in one of the development HS748s, coded H-2175, crashing in January 1999, killing key members of the Airawat team.

No. 50 Squadron, Indian Air Force.

Three A-50EIs serve with No. 50 Squadron at Air Force Station Agra.
(Georg Mader)

As a result fresh interest was shown by the IAF in the Beriev A-50, resulting in Russia loaning an example, which was flown for 60 hours. The hours were divided between 10 missions flown from Chandigarh in April 2000, manned by Russian personnel and accompanied by Indian experts. A contract for three A-50EIs was eventually signed by the IAF in 2004, these aircraft to be equipped with the Phalcon radar.

The first A-50EI was handed over on 25 May 2009, after overcoming difficulties integrating the Israeli radar and other equipment of Russian manufacture. The two remaining examples were delivered between April 2010 and February 2011, with the latter having improved and undisclosed features included. The actual radar chosen by the Indian Air Force was the ELW-2090, mounted in three radar antennas, each scanning a 120° arc, housed inside a fixed radome above the fuselage.

Indigenous solution

India still harboured the desire to develop an indigenous AEW&C platform and in 2003 issued a set of requirements for an aircraft that would complement the A-50EI fleet. The new aircraft would have to be able to fly for five hours when operating along the Indian-Pakistan border, be able to detect and track contacts with a minimum radar cross section of $5m^2$ at a range of 300km (186 miles), have the ability to guide fighters against them, and have at least a 120° arc scanning radar. The Embraer EMB-145 was duly selected as the new platform, and a contract signed by the Brazilian and Indian governments in July 2008. Under the agreement, Indian industry would be responsible for the mission equipment and systems, while Embraer would be responsible for the installation of them into the aircraft.

The EMB-145I ('I' standing for India) has a Primary Surveillance Radar allowing 270° coverage, using two antennas installed back-to-back in a dorsal radome. The data handling and display system comprises five workstations of the same type used aboard the A-50EI, four of which dealing with the interception missions, surveillance, communication and radar data.

A prototype, with a fixed in-flight refuelling probe (which allows the EMB-145I to double its mission duration) and dorsal antenna (without the radar equipment installed) was revealed in February 2011, and made its first flight on 7 December 2011. In June 2012 it flew to India, where installation of the on-board equipment began. This work was scheduled to take six months, after which the evaluation period would begin and will include at least 300 flying hours, after which it should be declared operational.

All three examples were scheduled to be inducted into IAF service in 2014, but at the time of writing the type was still undergoing testing with India's Defence Research and Development Organisation (DRDO).

The EMB-145I is part of an ambitious program announced in September 2012 and intended to give the IAF no less than five high-end large AEW&C aircraft which could include further A-50EIs or even a Western aircraft, such as the Boeing 767 or Airbus A330. As of August 2014, reports in the Russian media suggested India planned to sign a follow-on contract for three A-50s. India also hopes to field around 10 medium-sized AEW&C aircraft to augment its EMB-145Is, although the chosen type might not be the EMB-145I. It is planned that all types will include indigenous technology. The request for proposals for the new platforms was planned to be issued in 2014 and it is expected that the program will be concluded by 2020.

An EMB-145I displays the aerial refuelling probe found on the Indian Air Force aircraft. (Katsuhiko Tokunaga/D.A.C.T.)

Chapter 7

Indian naval AEW&C

In 1998 the Kamov Ka-31 entered low-priority production in the Kumertau Aircraft Manufacturing Factory, in the Russian Republic of Bashkiria. In August 1999 the type secured a first contract, when the Indian Navy ordered four examples, three of which were intended to serve aboard Krivak III class destroyers, with the remainder operating from the aircraft carrier *Viraat*. Shortly after, a further five units were ordered, with the first flight of an Indian Ka-31 occurring on 16 March 2001. They were inducted into Indian Navy service in 2003. At the time of writing, India was negotiating for five more Ka-31s. The Indian Navy Ka-31s have an indigenous electronics suite, manufactured by Bharat Electronics Limited.

INAS 339 'Falcons' at INAS Hansa, Goa, is responsible for the Indian Navy's Ka-31 fleet. (Gautam Datt)

Israel

Israel achieved remarkable success with its E-2C Group 0 Hawkeyes in the engagements over Lebanon in 1982 and went on to modify its Hawkeyes with indigenous design enhancements, fitting refuelling probes and augmenting and enhancing the radar using modifications designed and provided by IAI's Elta division. Even so, the E-2s was still short on endurance, a problem Israel was determined to solve in its next generation of AEW&C aircraft.

The problem was initially broached in the early 1990s by IAI and its partner companies Bedek Aviation (responsible for airframe modifications) and Elta Electronics (sensor systems) in the development of the Boeing 707-320C Phalcon that was equipped with an Elta EL/M-2075 electronically scanning phased array radar which was conformably mounted on the fuselage sides and is a bulbous nose radome. If 360° coverage is require then up to six active phased array scanner housings can be fitted in nose, cheek, wingtip and rear fuselage fairings. Alternatively, a triangular arrangement of antennas can be employed in pylon-mounted assembly above the fuselage.

The radar was one of the first in an aircraft to use the electronic scanned array system. There are no rotating antennas on such a system and the scan rate and direction is electronically controlled by the system computer to concentrate on selected areas in selected modes. The Phalcon system was original in fusing not only the raw radar contact data from the phased array but also integrated signals from the on-board IFF, ESM/ELINT and COMINT sensors and combining all of the data together to direct the beams of the radar to track targets. Additionally the data from ground radar stations can also be filtered into the system to create a whole 'battlespace' picture.

IAI was contracted by Chile to convert a Boeing 707-320C airliner into an AEW&C platform utilising the Phalcon system in 1991, the finished product was named Condór and was finally delivered to its owner in 1993. IAI has retained a second Boeing 707-344C airframe with the Israeli registration 4X-JYS and tail number 246, which is used by IAI as a development airframe for the Phalcon system and which was probably used to develop the enhanced system fitted to the Gulfstream G550 Conformal Airborne Early Warning and Control (CAEW) aircraft. Sources indicate that this aircraft was upgraded over a seven-month period in 2010 and flew again with the upgraded systems in January 2010 from Ben Gurion Airport, Tel Aviv. A third partially equipped Phalcon Boeing 707 is believed to have been converted for the South African Air Force (SAAF).

Serial number 569 is one of two G550 CAEW in service with the IASF. Interestingly, it carries no national insignia on the wings. (Dr Andreas Zeitler)

No. 122 'Nachshon' Squadron, Israel Air and Space Force.

Nachshon Eitam

In 2001 Israel started the process of reinforcing and renewing its fleet of special missions aircraft, and chose the Gulfstream G550 airframe as a basis for versions to adapted for battlefield surveillance and electronic warfare. In the second half of 2003, it was announced that Gulfstream had been awarded a USD473-million contract for four aircraft plus an option for two more that were to be used in Israel's Conformal Airborne Early Warning and Control (CAEW) programme.

The new G550 Nachshon Eitam (pioneer – fishing eagle) was fitted with the EL/W-2085 radar that was an evolution of the EL/W-2075 and EL/I-2080, which had originally been installed in the IAI Condór. It comprises of two pairs of antennas: the first pair are conformal scanners mounted on the forward fuselage with a scanning angle of 135° each, while the nose and tail scanners cover 40° and 50° respectively. This 360° radar coverage works in conjunction with ESM/ELINT systems (located on the wingtips and in the radomes), and there are also ESM/COMINT and self-protection suites located under the engine nacelles.

The fuselage interior of the G550 has its forward section occupied by electronic equipment racks, while the rear portion contains six positions for equipment operators. The airframe has been adapted with enhanced liquid equipment cooling and an up-rated power generator. The communications suite provides network-centric operations capability and includes U/VHF, HF, and satellite communications, voice over internet protocol (VoIP) and a secure data-link.

The CAEW prototype registered N637GA first flew in May 2006 in the US and was flown to Israel on 19 September 2006 for the installation of the mission avionics, while the last example, N969GA, was ferried for modification in July 2007. The type was officially declared operational in February 2008 and received the Hebrew name Nachshon Eitam. The aircraft was in fact flown on operational sorties prior to being declared operational. With the retirement of its E-2Cs in 2002, Israel had lacked an AEW&C asset for five years, and the Nachshon Eitam was probably employed in Operation Orchard, the attack on Syrian nuclear facilities by Israeli F-15Is and F-16Is, in September 2007. Additionally, two Nachshon Eitams were flown during the early stages of Operation Cast Lead in Gaza between December 2008 and January 2009 when they precisely located ground targets and generated a complete picture of the operational theatre for Israeli commanders. The Israel Air and Space Force's No. 122 Squadron operates two Nachshon Eitams (serial numbers 537 and 569).

Italy

An AEW version of the AgustaWestland AW101 Merlin resulted from a requirement issued by the Italian Navy in 1994 for a platform dedicated to the heliborne early warning role from both onshore bases and ships, thus providing AEW coverage to the Italian fleet. Four examples were ordered the following year. While the Italian military uses the service designation EH-101A, the manufacturer designates the aircraft as the Merlin Mk 112.

The first aircraft to fly after conversion to the AEW role was serial number MM81488 on 1 December 1999 and the delivery of all four helicopters to the Italian Navy was completed in 2006. The AEW Merlins are based at La Spezia with GRUPELICOT 1 and are compatible with Italian Navy ships including the aircraft carrier *Cavour*.

In 2012 Israel Aerospace Industries signed a deal with Rome to supply the Italian Air Force with two Gulfstream G550-based Conformal Airborne Early Warning (CAEW) and control system aircraft, worth USD750 million. Also included are ground support equipment and logistics services. The contract was part of a counter-deal to Israel's USD1-billion order for 30 Alenia Aermacchi M-346 advanced jet trainers. The first G550 CAEW for the Italian Air Force was expected to be delivered in late 2015.

The interior of an EH-101A AEW helicopter.
(Marina Militare via Massimo Tiberi)

An Italian Navy EH-101A operating from from Sarzana, home of 1° Gruppo Elicotteri that also operates the SH-101A (anti-submarine/anti-surface warfare) version of the Merlin.
(Luca La Cavera)

Japan

Japan operated dual-role AEW/ASW TBM-3W2 Avengers between the mid-1950s and early 1960s but they were retired without replacement. Alternatives to meet an emerging AEW&C requirement were examined by the then Japan Defense Agency (later Japan Ministry of Defense), which initially selected the E-3A Sentry. At the time, the E-3 was still in development, with the first deliveries to the USAF scheduled to take place the following year. Other reasons against its selection were its maximum all-up

Since 2014 Japan has operated two E-2C squadrons, 601 and 603 Hikotai, due to increased Chinese activities. (Katsuhiko Tokunaga/D.A.C.T.)

601 Hikotai, Japan Air-Self Defense Force.

weight, which would require the available Japanese runways to be reinforced, its high acquisition price and the fact that its control capability was judged superfluous.

In particular, the need for an AEW&C platform was exposed in September 1976, when Soviet pilot Viktor Belenko defected to Japan with his Mikoyan-Gurevich MiG-25P interceptor. After managing to evade the F-4Es of the Japan Air Self-Defense Force scrambled to intercept him, Belenko went unnoticed by radar ground stations until he suddenly popped up over Hakodate airport and carried out a forced landing.

To fulfil the urgent need for an AEW asset, 11 E-2C Group 0 Hawkeyes were acquired in two discrete batches, in 1982 and 1993.

E-767

Defence plans published in 1991 and 1992 envisaged the acquisition of a platform that would have data-gathering, data-link and control capabilities and be able to loiter above the sea and airspace lanes around the Japanese archipelago for a much greater length of time than the Hawkeye. AEW&C versions of the C-130 and P-3 were considered, but the Japanese government did not want another turboprop model to complement the E-2s. The obvious choice was the E-3. Unfortunately, the assembly line had been closed in 1991, but as an alternative Boeing suggested a modified version of its 767-200ER, designated as the 767-27C/27CER, later E-767. The proposal was accepted and in November 1993 Boeing was awarded a USD840-million contract from the US Air Force for the integration/installation of mission equipment aboard two 767s, followed by a further USD773 million for another two aircraft, 11 months later, all of them to be acquired via a Foreign Military Sales agreement.

The basic 767 airframe was manufactured by Boeing in Everett, Washington, and then flown to the Boeing facilities in Wichita, Kansas, where the airframe was modified to accommodate the prime mission equipment. All aircraft were then returned to Boeing's Seattle facility where mission equipment, including the rotodome, was installed and the flight testing took place. Major subcontractors included Northrop Grumman,

Chapter 7

A JASDF E-767 of the Hiko Keikai Kanseita (Air Warning Control Squadron) of the Keikai Kokutai (Airborne Early Warning Group) based at Hamamatsu. In 2014 the unit was renamed 602 Hikotai.
(Katsuhiko Tokunaga/D.A.C.T.)

General Electric, Rockwell Collins and Telephonics, which had all been involved in the previous Boeing AWACS programs. Most of the mission equipment was installed internally in the forward fuselage, counterbalancing the weight represented by the rotodome and crew rest area (which also includes a galley and a lavatory) in the rear of the aircraft. The control and air surveillance area is concentrated on the right side of the aircraft and comprises two back-to-back rows of four workstations with dual monitors on each.

The E-767's operational features are very similar to those of the E-3C Block 30/35. The radar is the Westinghouse AN/APY-2 (with maritime scan-to-scan processing, low-velocity detection, pulse-Doppler assessment and ESM blanking pulse output), with

The cavernous interior of the E-767, showing the radar operators' consoles, and plenty of room for future expansion.
(via Marcia Costler)

data analysis performed by a 4Pi CC-2E computer which includes the NATO expanded memory upgrade and control power supply including connectors, backplane, wiring, and space for future addition of an ESM interface module.

The first totally modified aircraft flew in October 1994 and the first flight with the rotodome installed occurred on 9 August 1996. The initial E-767 (a -27C subtype, registered N767JC, with the JASDF serial number 74-3503) was delivered on 12 June 1997, followed by 74-5304 on 15 January 1998. The third and fourth E-767s – 64-3501 and 64-3502 – were inducted to service on 1 May 1998. On 1 April 2000 the type reached full operational capability. The aircraft were assigned to the Hiko Keikai Kanseita (Air Warning Control Squadron) of the Keikai Kokutai (Airborne Early Warning Group) based at Hamamatsu. In 2014 the unit changed its name, becoming 602 Hikotai.

In December 2010, Boeing won a contract to install and test the RSIP and a mission navigation system upgrade. This work was completed in December 2012. The RSIP provided rewritten software, improved radar sensitivity, and a more reliable computer with multi-processors that allow the E-767 to detect and track cruise missile-sized targets. Another modernisation programme, the Mission Computing Upgrade (MCU) is intended to enhance the E-767's command and control capability, and was announced in late September 2013. The MCU involves the purchasing of four ESM of an undisclosed type, eight AN/UPX-40 Next Generation Identify Friend or Foe (NGIFF) systems, eight AN/APX-119 IFF transponders, and four KIV-77 cryptographic computers.

In summer 2014 the Japan Ministry of Defense announced defence plans that included a request for USD531 million to spend on four new AEW&C aircraft by 2018, with a choice to be made between the E-2D or Boeing 737 AEW&C. At the same time, defence budget requests for 2015 were expected to include USD125 million for upgrades to the E-767 fleet. The Ministry of Defense is also examining the possibility of manufacturing an indigenous AEW&C aircraft by the mid-2020s. Such a platform would be procured in addition to the US-made successor to the E-2C. Should the indigenous programme receive funding, construction of a prototype will begin on the basis of the Kawsaki P-1 maritime patrol aircraft. The AEW&C derivative would use radar technology based on ground-based radar carried in a fuselage-mounted rotodome.

Mexico

One of three Mexican Navy E-2Cs, serial number AMP-102 made a rare visit to the UK in November 2004. The type was mainly used to protect Mexico's maritime oil reserves. (David Unsworth)

The Mexican Air Force's sole EMB-145AEW&C. (José Antõnio Quevedo)

The Mexican Navy (Armada de México) operated three ex-Israeli E-2Cs, the first two of which entered service on 1 July 2004. Operated by the Primer Escuadron Alerta Temprana y Reconocimiento (PRIMESCATREC) at Tapachula naval air base, the aircraft were primarily used for what was termed 'deep maritime patrol' (including AEW) over Mexico's offshore oil installations, mainly in the Gulf of Mexico. The aircraft's service lives ended in July 2007. While one remains in open storage the other two have been scrapped.

In 1999 the Mexican Air Force (Fuerza Aérea Mexicana, FAM) announced a requirement for an AEW&C aircraft, to enhance the fight against drug trafficking and strengthen the military capability to respond to natural disasters. A contract was signed with Embraer for a single EMB-145AEW&C. The contract included spares, ground support and staff training, and training of aircrew in regards to electronic surveillance, air traffic control and detection, tracking, interception and capture of air and sea targets suspected of illegal activities. pilots, radar operators and ground crews were trained in Brazil, their instruction period concluding in 2002 to coincide with the delivery of the first aircraft.

On 16 December, 2004, the Sistema Integrado de Vigilância Aérea (SIVA, Aerial Surveillance Integrated System) was activated, and comprised a Command and Control Centre, Aerial Surveillance Squadron, Detection and Control Group No. 1, Simulation and Training Centre, and Advanced Maintenance Workshop. The single EMB-145AEW&C is operated by the Escuadron de Vigilancia Aerea that forms part of SIVA and which is split into three flights (escuadrillas) each with a variety of aircraft.

Mexico's EMB-145AEW&C is similar to the model flown by Brazil, but its PS-890 Erieye radar is capable of 360° coverage. The main mission of the single EMB-145AEW&C is to detect, track and identify targets in its patrol area and relay that information to friendly forces. In addition to its AEW&C capabilities the type can also perform ELINT missions. With the operation of this aircraft the FAM has the capability to help to prevent or to coordinate a response to an attack by air or sea from terrorism on strategic facilities.

Mexico intends to acquire a second AEW&C asset by 2018, although the type has yet to be confirmed.

NATO

With new upgrades, NATO E-3s will soldier on until at least 2024, albeit in smaller numbers. (Dr Andreas Zeitler)

NATO became interested in AEW platforms in the early 1960s, when its air forces began operating at low altitudes, and its commanders realised that ground-based radars were unable to detect and track such aircraft. After attempting to solve the problem by increasing the number of early warning radar stations from 18 to 40 between 1966 and 1972, the emergence of the MiG-27 fighter-bomber in Soviet service in the mid-1970s forced the NATO's Supreme High Command to re-think its strategy. Such fast and low-flying fighter-bombers were capable of reaching most of West Germany entirely undetected while flying at low altitude. However, if NATO were to acquire an AEW platform that could orbit at 9,144m (30,000ft) above the Inner German Border, it could provide at least half an hour of warning time for NATO interceptors to scramble.

Furthermore, NATO commanders wanted to obtain a maritime patrol capability to protect their crucially important communications to the US, and provide early warning of any possible amphibious assault by the Warsaw Pact in the Baltic. NATO approached Boeing with a request to adapt the Sentry into a multi-purpose platform that could meet its requirements and, despite British objections, based on London's insistence on the stillborn Nimrod AEW.Mk 3, signed a corresponding contract in 1977. Although Boeing's offer price for E-3s to NATO was cheaper than that for the USAF, all of the then European NATO members (except the UK) contributed to the cost of purchasing AWACS from the US. In May 1978, after nearly four years of negotiations, the European defence ministers finally signed an order for 18 aircraft worth a total of USD1.9 billion, shared proportionally between the US, West Germany, Canada, Italy, Belgium, Denmark, Norway, Turkey, Luxemburg, Portugal and Greece.

The purchase agreement for the NATO E-3As stipulated that Boeing was to only install radar and navigation systems into the airframes, while much of the remainder of the avionics fit would be produced by different European companies. For example, SEL of West Germany produced communications and display screens; ESG, MTU and AEG-Telefunken manufactured many other airframe and avionics components. Furthermore, the addition of the maritime surveillance capability resulted in the emergence of a new variant of the main radar, the AN/APY-2, which proved so successful in service that it was retrofitted into USAF E-3As.

The NATO Airborne Early Warning & Control Force (NAEW&CF) was established in January 1980. The NATO E-3A main operating base (MOB) is at Geilenkirchen, Germany. Forward operating bases (FOBs) are located at Trapani, Italy; Aktion, Greece;

Chapter 7

and Konya, Turkey. There is also a forward operating location (FOL) at Ørland, Norway.

In the mid-1980s, the Block 10 Sentries that had been sold to NATO were modified to Block 25 configuration that added five UHF radios. Between 1992 and 1999, the NATO Sentries were further modernised, which involved the incorporation of Link 16 data-link, an interference-proof UHF communication system, colour console screens, ESM and extended radar memory, all of which were in accordance with the standard USAF Radar System Improvement Program (RSIP) upgrade.

In August 2014 Boeing announced receipt of a USD250-million contract to upgrade NATO's E-3A fleet. Thirteen aircraft will receive new avionics designed to meet compliance with future aircraft regulations. Modifications will take place between 2016 and 2018. The upgrades will also reduce the number of flight crew from four to three. With the latest upgrade applying to 13 aircraft, the alliance is conducting a force review that will likely see the three remaining aircraft retired. Another aircraft is receiving the flight deck upgrade as part of the engineering, manufacturing and development effort. In line with the latest modernisation work, the NATO E-3 fleet is expected to serve until at least 2024, and perhaps until 2035.

NATO E-3As are operated by international crews from 16 nations (Belgium, Canada, Czech Republic, Denmark, Germany, Greece, Hungary, Italy, the Netherlands, Norway, Poland, Portugal, Romania, Spain, Turkey and the United States).

Pakistan

The Pakistan Air Force (PAF) identified an operational need for an AEW&C asset during the 1980s, when the Soviet Union invaded Afghanistan. This requirement was only fulfilled in the fourth quarter of 2005 when the Pakistani government signed a contract with Saab for the development of an AEW&C version of the Saab 2000 regional airliner. Initially six units were required but this was reduced to five in May 2007, including one dedicated crew trainer. The airframes were drawn from the Saab leasing stocks and were zero-houred and re-engined.

External modifications included the installation of the Erieye radar, communication antennas, the elimination of nearly all its starboard-side cabin doors, the instal-

One of Pakistan's Saab 2000 AEW&Cs at the Dubai Airshow in November 2001.
(Katsuhiko Tokunaga/D.A.C.T.)

Pakistan's ZDK-03 complements the Saab 2000 AEW&C, however, its radar reportedly offers greater range than the Erieye and its avionics incorporate a greater degree of open architecture. (Shawn Chen)

No. 4 Squadron, Pakistan Air Force.

lation of wingtip ESM housings, fins on the vertical stabiliser which was extended to 8m (26ft), and small vortex generators on the wings and tail. Internally, the aircraft is manned by two pilots, behind which there is a lavatory, followed by a rest area with six seats and a table, and a mission compartment with five operators' positions with touch input/high-definition colour displays. The resulting platform is considered to be able to make 50° bank angle turns in 30 seconds and accelerations of $1.7g$.

The first example left the factory in April 2008 and was delivered to No. 4 Squadron, Pakistan Air Force in December the following year, with the second being passed to the PAF in April 2010. In August 2012 at least one of the aircraft was destroyed by a terrorist attack against Kamra Air Base.

Chinese AEW for the PAF

China began development of the ZDK-03 in the early 2000s as an AEW&C system dedicated to the export market, and based on the Y-8F-600 airframe. It was offered to Pakistan in 2006, and four examples were ordered in 2009. The first aircraft, serial number 11-0001, was delivered to Pakistan on 26 November 2011. Full acceptance checks on the aircraft, also known as the Karakoram Eagle, were completed by 29 December 2011.

Qatar

In a surprise March 2014 announcement, the Qatar Emiri Air Force (QEAF) emerged as the latest customer for the Boeing 737 AEW&C. Under a USD24-billion deal, Qatar will receive three 737 AEW&Cs amounting to USD1.8 billion of the total procurement cost. A timeline for the delivery of the aircraft has not been revealed.

Russia

The Russian Air Force's fleet of A-50s, around a dozen of which remain in use with one squadron based at Ivanovo, northeast of Moscow, is subject to modernisation by Beriev. The A-50U (oosovershenstvovannyy – upgraded) standard primarily features new electronics and can be distinguished from the basic A-50 (as described in Chapter 4) by the deletion of the large, lateral strakes on each side of the rear fuselage.

Russian Air Force A-50U 'Red 33' is the sole example to have received this grey 'NATO-style' paint scheme. (Sergey Krivchikov)

Confusingly, the A-50U designation was also applied to a less extensive upgrade dating from the late 1980s. A first A-50U ('Red 47') was delivered to the Russian Air Force in October 2011 for operational testing and official service entry was declared in January 2012. The latest upgrade features digital avionics, more powerful computers for faster data processing, improvements to the radar and tracking capability (especially in the rear hemisphere), improved communications and navigation equipment, modernised work stations with LCDs replacing analogue displays, and new crew rest areas, galley and toilet facilities. With the introduction of lighter equipment, fuel capacity is increased, extending range and endurance. The third A-50U ('Red 53') was returned to the Russian Air Force at Taganrog in March 2014. The aircraft is the first A-50 to have received an individual name, *Sergei Atayants*, in honour of its chief designer.

In 2000 it was proposed that the A-50 could be further developed and enhanced for the Russian Air Force along similar lines to the versions upgraded for China and India. It was proposed that a new variant would be equipped with phased array antennas housed in a new, fixed radome.

At first, NGO Vega to be the main contractor, but financial difficulties precluded its involvement and it was substituted by MNIIP Tikhomirov, which expanded the initial proposal to include the ability for the new aircraft to be used as an intelligence platform. Two years later, the project was officially presented and in 2003 it was awarded a contract by the Russian Defence Ministry for the further development of the system, to be executed by Vega and Tikhomirov. In late April 2004, a Presidential decree authorised the system, which was to be installed in the updated Il-76MD-90A, or perhaps even aboard the massive Antonov An-124 cargo aircraft. The system was designated A-100, and the Il-76MD-90A was chosen as the platform for the new system, of which the main component is the Vega Premier rotating phased-array radar. The Premier employs mechanical scanning in azimuth and active electronic scanning in elevation and rotates twice as fast as the previous Shmel radar. As of August 2014, experimental on-board equipment for the new aircraft had been completed and its design documentation prepared, paving the way for acceptance trials of the equipment in 2016. The Beriev A-100 could begin acceptance trials in 2017, according to Vega officials.

The main mission operator compartment of the A-50U. (Tver News)

Ka-31 'Blue 032'.
(Piotr Butowski)

Russian naval AEW

The Ka-31 was evolved from the Kamov Ka-29 assault transport helicopter, and first flew on 25 November 1986. The Ka-31 emerged as the only practical solution to meet the AEW&C requirements of future Soviet Navy aircraft carriers, after the failure of the fixed-wing Yakovlev Yak-44E and Beriev P-42 projects.

The conversion work for the new variant started in 1985, under the supervision of the Deputy Chief Project S. N. Fomin, resulting in an electronic mission suite arranged around the E-801M Oko (eye) radar, whose 6 x 1m (19.7 x 3.3ft) planar antenna. The radar is folded and stowed beneath the aircraft's fuselage before being lowered into a vertical position, to allow 360° mechanical scanning of the radar once every 10 seconds. The radar installation required an increase in width from 1.4m (4.5ft) to 2.41m (7.9ft) between the front landing gear legs. The second crewmember lowers the antenna, activating it, selects the scanning mode and supervises the operation of the system, with detected contacts being displayed on a 15 x 20cm (5.9 x 7.8in) touchscreen display with the synthetic image of the target trajectory and its data, combined with a digitised map of the scene.

During tests of the electronic equipment between 1991 and 1995, it was found that the system extended the radar coverage of a task force from 10-15km (6.2-5 miles) to 300-350km (186-217 miles).

In November 2008 Russian Naval Aviation signed a contract to receive two Ka-31R versions. The initial example of the pair, 'Red 90', was transferred to the Naval Aviation training centre at Eysk in June 2012, and has been followed by 'Red 91'. These were delivered to the 859th Evaluation Centre in Yeysk. In April 2012 a tender was announced for one further Ka-31R to be delivered by 25 November 2013, but no offer was submitted and the tender was cancelled. As a result, as of September 2014, just two production-standard Ka-31Rs are active in Russia, and these may have been returned to the Kamov Design Bureau.

Chapter 7

Saudi Arabia

Following hard on heels of Iran's attempt to purchase the E-3 was a delegation from the Royal Saudi Air Force (RSAF). Saudi Arabia had been protected by the US military since the 1940s but by the mid-1970s it began making enough profits from oil sales to afford to commence a major expansion of its military. This included the acquisition of F-15 interceptors and establishment of a ground-based integrated air defence system. Saudi Arabia did not have a problem with mountain chains, but the size of the country and its lengthy borders were such that they dictated the acquisition of an AEW aircraft. Saudi demands for E-3As caused another major controversy, as a result of Israeli objections: not only did Israel see Saudi E-3s as a major threat to its own security, but their appearance could have strongly hindered the freedom of operation of the IDF/AF in regards to future attacks on Iraq, as they had in June 1981, when Israeli jets used the airspace over northern Saudi Arabia to enter Iraq.

After much cajoling, extensive negotiations and the provision of security guarantees for Israel, a contract was signed with Riyadh in February 1980, launching Project Peace Sentinel. This approved deliveries of five E-3As and eight KE-3A tankers, starting in 1984 with the construction of the purpose-built Prince Sultan AB near al-Kharj, 80km (50 miles) southeast of Riyadh, and the commencement of training of personnel of 18 Squadron RSAF. Although the contract was signed and preparations for its fulfilment launched, the final 'green light' for the sale of the E-3s to the Saudis was not finally authorised by US President Ronald Reagan until October 1981.

Interestingly, the RSAF (like the French and the British) opted to power its E-3s with the highly efficient CFM56 high-pass turbofan engines, making them more suitable to operate in the hot climate and orbit at higher altitudes.

In August 2014 the US State Department approved a modernisation program for the RSAF Sentry fleet, to be provided in the form of a Foreign Military Sale worth USD2 billion an also including equipment, parts, training and logistical support for an estimated. Under the proposal, all five E-3As would be brought up to Block 40/45 standard via the Mission Computing Upgrade, and would also receive AN/UPX-40 Next-Generation Identification Friend or Foe (NGIFF) systems, and new communication equipment.

The latest addition to the Saudi AEW&C fleet is a single Erieye-equipped Saab 2000 AEW&C, serial number 6601, operated by 66 Squadron at an unknown air base.

No. 18 Squadron, Royal Saudi Air Force.

Saudi E-3s can be seen in the UK when on their way to and from the US for regular overhauls. RSAF E-3A serial number 1803 was a visitor to RAF Mildenhall in May 2014. (Duncan Monk)

Delivery of the aircraft was reported in October 2010, when a cost of USD669 million was quoted. The aircraft was acquired together with ground equipment and logistics and support services. In March 2013 Saab announced a follow-on support contract for a single Saab 2000 and Erieye, operated by an undisclosed customer, and valued at USD170 million, to run until 2017.

Singapore

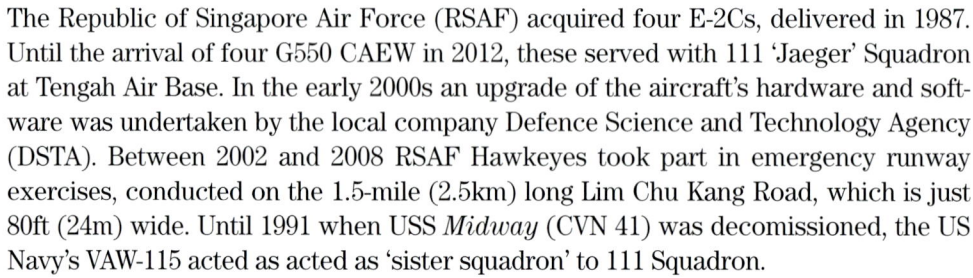

The Republic of Singapore Air Force (RSAF) acquired four E-2Cs, delivered in 1987. Until the arrival of four G550 CAEW in 2012, these served with 111 'Jaeger' Squadron at Tengah Air Base. In the early 2000s an upgrade of the aircraft's hardware and software was undertaken by the local company Defence Science and Technology Agency (DSTA). Between 2002 and 2008 RSAF Hawkeyes took part in emergency runway exercises, conducted on the 1.5-mile (2.5km) long Lim Chu Kang Road, which is just 80ft (24m) wide. Until 1991 when USS *Midway* (CVN 41) was decommissioned, the US Navy's VAW-115 acted as acted as 'sister squadron' to 111 Squadron.

Singapore's four G550 CAEW aircraft were purchased from Israel in February 2009 at a reported cost of USD1 billion. The G550 CAEWs are operated by 111 Squadron at Tengah and provide an endurance of around nine hours as compared to the Hawkeye's four hours. Like Israel's Eitam, the G550 AEW has six controller positions and is equipped with a line-of-sight data-link and satellite communications. The final aircraft of the batch was delivered in October 2011 and 111 Squadron was declared fully operational on 14 April 2012. Singapore's four G550 CAEWs comprise serial numbers 010, 016, 017, 018, although prior to delivery these wore fake Israeli serial numbers while being tested at IAI.

111 'Jaeger' Squadron, Republic of Singapore Air Force.

A Republic of Singapore Air Force E-2C prepares to launch from a highway during an alternate runway exercise in November 2008.
(Ow Eng Tiong/Alert 5)

RSAF G550 CAEW serial number 017 participated in Exercise Pitch Black 2014 at RAAF Darwin, Australia.
(Ow Eng Tiong/Alert 5)

Chapter 7

South Korea

The Republic of Korea Air Force (RoKAF) first identified a requirement for AEW&C assets in 1980, when it became clear the country's topography would impose serious limitations on the detection performance of ground-based radar stations. However, an initiative to overcome these difficulties only appeared in 2005 when the E-X programme was launched. The initial list of competitors included the already described G550 CAEW, promoted by an American-Israeli consortium of General Dynamics Gulfstream, L-3, and IAI Elta and the Boeing 737 AEW&C. The former was the preferred option, mainly due to its low price and operating costs, but the RoKAF was forced to turn down the G550 CAEW because some of its systems, such as Link 11 and Link 16 data-links, IFF, SATCOM module, UHF/VHF Have Quick radios and GPS P(Y) code technology were not available for export to South Korea. Boeing's 737 AEW&C became the sole competitor and it was officially declared winner in August 2006, after an eight-month delay in the process. In order to avoid similar restrictions in the future, the restricted hardware was replaced, where possible, by indigenous counterparts.

Four examples of the 737 AEW&C were ordered under a USD1.6-billion Peace Eye programme that commenced in 2006. Although the first example was constructed by Boeing in Seattle, Korean Aerospace Industries (KAI) performed aircraft modification and mission equipment installation on the remaining three aircraft in Saechon, South Korea. Three of the four aircraft were delivered to South Korea between February 2010 and May 2012. The fourth and final aircraft, serial number 65-329, was delivered on 24 October 2012.

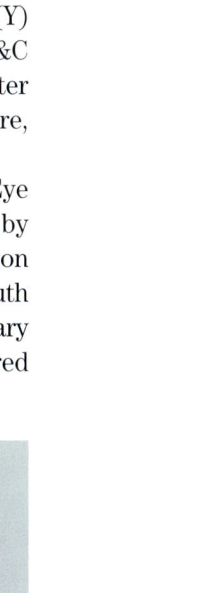

271st AEW&C Flight Squadron, Republic of Korea Air Force.

The RoKAF 737 AEW&C fleet serves with the 271st AEW&C Flight Squadron at Gimhae Air Base.
(Taegu Lee)

Spain

Following hard on the heels of the RN, the Spanish Navy purchased four Searchwater sets and installed them in three of its own Sikorsky SH-3H Sea Kings, re-designating them as SH-3(AEW)s, and in one ground simulator. The helicopters became operational from the carrier *Príncipe de Asturias* in 1989 and today two airworthy specimens are available to operate from the carrier *Juan Carlos I*.

Spanish Navy Sea Kings are operated by the Quinta Escuadrilla (Fifth Squadron). The unit includes the two remaining operational SH-3(AEW)s, based at Rota.
(Erik Roelofs)

In May 2011 it was revealed that Airbus Military and IAI Elta were jointly developing an AEW&C version of the former's C295 turboprop cargo aircraft, targeted at nations that require an AEW&C asset but have limited defence budgets. The main radar will be a 360° AESA type that is still under development by Elta and will be housed within a rotodome. The data gathered will be managed by the Airbus Fully Integrated Tactical System (FITS) to be installed in six workstations surrounded by systems cabinets and a rest area. In addition to being capable of high-priority target surveillance, the radar can remain fixed to provide highly accurate tracking of targets within a specified 120° sector. The system also has SIGINT and self-protection systems which integrate signals received from RWR, MAWS, ESM/ELINT and CFD. The aircraft is said to have a time on station of up to eight hours and a radar range of 370km (230 miles). As of September 2014, no customers for the C295 AEW&C had been announced.

Sweden

72. Ledningsflygdivisionen, Swedish Air Force.

In 1977 the Swedish Air Defence for the Next Century study suggested the indigenous development of a radar system that was compact enough to be able to be accommodated in a fuselage compartment of an aircraft type much smaller than those normally used for early warning and control. In autumn 1985 Ericsson AB received a directive from the Swedish government to begin studies into a system to be known as Pulse Surveillance 890 (PS-890) that would be promoted commercially as Erieye – the 'eye of Ericsson'. A Fairchild Swearingen Metro III (Swedish military designation Tp88C) serial number 883 was chosen as airborne test bed for the programme in spring 1986. In September 1987 the Saab 340B twin turboprop was selected as the primary carrier aircraft of the Erieye. To compensate for the aerodynamic effect of the 9.7m (31ft 9in) long 'balance beam' antenna pylon-mounted above the fuselage, two strakes were added to the lower rear fuselage to improve lateral stability and attenuate some of turbulence created around the vertical fin.

The first flight with the antenna installed occurred on 1 July 1994, and the first aircraft, now with the military designation S100B Argus, was delivered to the Swedish Air

Chapter 7

Serial number 100004, one of two Saab S100D Argus operated by the Swedish Air Force. (Dr Andreas Zeitler)

The Saab ASC 890 mission compartment, looking forward. (André Caldenius)

Force (Svenska Flygvapnet) in 1997. All six fully operational aircraft were delivered to the Swedish Air Force by November 1998, with the complete system now bearing the service name FSR 890 (Flygburen Spaningsradar, airborne surveillance radar). The cruising speed while in operation is kept as low as possible to save fuel and extend the sortie duration, and this slow speed requires that the aircraft is flown at a nose-up attitude of three degrees. This in turn dictates that the antenna is set at a corresponding slope angle in relation to the horizon. The Erieye radar is not able to detect contacts to the front and rear of the aircraft as its beams cover a look angle of only 160° either side of the aircraft. The Erieye is able to select different pulse-Doppler rates depending on the type of target being tracked: medium PRF for air targets and low PRF for naval targets. The radar is an active-array type with 2,000 solid-state modules and operates in the S band. The antenna is composed of two rows of modules positioned vertically and divided into upper and lower portions. Using oscillation control within the modules enables the operators to ascertain the altitude of detected targets.

When originally introduced, the Swedish Air Force's Argus did not carry radar operators aboard the aircraft. An on-board radar technician monitored the systems, but the picture would be down-linked via NATO-compatible data-links to radar operators position at radar consoles and ground stations. When orbiting, normally in racetrack pattern legs of up to 100km (62 miles), the side of the radar facing away from the threat direction is switched to standby. As the aircraft turns at the end of the leg the standby side is turned on and the active side reverts to standby; in this way power use is decreased, heat build-up is diminished and full power can be used on the direction of interest. As the sweep is generated, the radar operators can select areas of special interest for the beams to be concentrated against and select which mode of scan is required for each particular type of target.

In 1990 the Swedish Air Force upgraded its air defence systems to STRIL 90 standard that was designed to modernise all command, control, communications and intelligence systems. Thus, AEW platforms and fast jets could operate if required in a data-link-only environment without voice communications. The connection between the mission system on the aircraft and the ground command cell occurs through the FR-JAS device that comprises three V/UHF radios.

In the 1990s only four of the six Argus aircraft were equipped with the Erieye radar actually installed and functioning and the remaining two were used as transport and communications aircraft.

When the two Saab 340Hs loaned to Greece were returned to Sweden, the operational experience gained by them was not wasted. In the second half of 2006 the Swedish government launched a programme to provide a new AEW&C platform, based on the Greek Interim Solution and capable of acting as part of an international force. Saab Aerotech and Ericsson Microwave were awarded two contracts to expand the mission avionics on the Saab 340Hs. The modification process of the two aircraft lasted until 2007, and the following year the S100D Argus aircraft were delivered to the Swedish Agency for Defence Equipment (FMV, Forsvarets Materiel Verk) for evaluation. They were handed over to the Swedish Air Force in mid-2009 and achieved initial operational capability in 2010. The system now carries the revised designation ASC 890 (Airborn Surveillance and Control). The operating unit is the 72. Ledningsflygdivisionen, part of F 7 Såtenäs wing based at Såtenäs/Malmslätt.

Taiwan

In May 1993, Taiwan signed a contract with Grumman to supply four E-2T Hawkeyes (the T standing for Taiwan) at a cost of USD700 million. The original aircraft supplied were rebuilt E-2Bs fitted with AN/APS-138 radars, and the first example was delivered to the Republic of China Air Force (RoCAF) on 9 May 1995. Two others were delivered by sea and arrived on 3 September of the same year, followed by the second pair on 23 September. Even though all five had E-2B Bureau Numbers, RoCAF officials claimed that they were in fact all new-build Hawkeyes built to E-2C Group II standard and fitted with the more advanced AN/APS-145 radar. It was also said that a small number of parts had in fact been recycled from some old E-2Bs. In another USD400-million deal, the sale of a further two E-2Ts (upgraded to Hawkeye 2000 standard) was announced by the US Department of Defense on 31 July 1999 along with a large quantity of avionics spares and a support programme. The first of the two Hawkeye 2000s was accepted by RoCAF officials at Northrop Grumman's St Augustine, Florida facility on 10 August 2004. The second was received in October 2004. Again both were transported to Taiwan by sea.

A RoCAF E-2 Hawkeye 2000 from the 2nd Early Warning Squadron, based at Pingdong South Air Base. (Dr Andreas Zeitler)

Thailand

In the fourth quarter of 2007 it was announced that the Royal Thai Air Force would sign two separate agreements to acquire 12 Saab JAS 39C/D Gripen fighters and two S100Bs plus a standard Saab 340 transport to act as a pilot trainer and utility aircraft. All the aircraft were delivered between 2010 and 2012 and serve with 702 Squadron at Wing 7, Surat Thani RTAFB.

A Saab 340 AEW&C patrolling Thai coastal waters. The aircraft serves with 702 Squadron at Surat Thani RTAFB. (Katsuhiko Tokunaga/D.A.C.T.)

Turkey

As was the case with South Korea, the Turkish Air Force (Türk Hava Kuvvetleri, THK) also concluded in the late 1990s that the geographical features typical of Turkish territory would prove difficult for ground-based radars to operate efficiently and pick up low-flying, fast-moving aircraft. It was decided to acquire an organic AEW&C capability to fill the gaps, and two proposals were put to the THK, comprising the Airbus A310/Phalcon radar combination and the Boeing 737 AEW&C. In late December 2001 the

Turkish Air Force 737 AEW&C serial number 10-001 participating in Exercise Anatolian Eagle at Konya air base in June 2014. (Hans Rolink)

131 Filo, Turkish Air Force.

latter was announced as the preferred option, and six aircraft were ordered at first. The number was reduced to four aircraft when the USD1.6-billion contract for the resulting Peace Eagle programme was finally signed on 23 July 2000. The first aircraft would be assembled by Boeing, while the remaining three were to be constructed by the Ankara-based Tusas Aerospace Industries (TAI). Subcontractors included Turkish Airlines for flight crew training, Havelsan of Turkey was to provide the ground installations and Elta of Israel the ESM equipment. According to the schedule of the contract the first aircraft should be have been delivered in July 2007, while the date for the delivery of the final aircraft should have been June 2008. However, the programme suffered various delays, one of them caused by an Israeli embargo on Elta exporting the ESM system.

The first aircraft to fly on 6 September 2007 was N356BJ, serial number 001, from Seattle; the first aircraft modified by TAI flew in July 2008. A further delay was caused by tensions between Israel and Turkey in the aftermath of the Gaza flotilla raid in May 2010.

This had consequences on the Peace Eagle project, which were only resolved in early 2013. The first aircraft was then scheduled to be delivered by the end of 2013, but actually arrived in Turkey in January 2014, six years late. In February 2014 Boeing finally announced that it intended to complete deliveries in 2015.

United Arab Emirates

In the last few months of 2009 Saab announced that the United Arab Emirates Air Force would acquire two S100Bs as an interim solution for a future requirement for AEW&C assets. The aircraft were delivered between late 2010 and early 2011 and have had a new IFF fitted.

A rare glimpse of one of the UAE's Saab 340 AEW&C aircraft during a 2013 visit to Germany. (Dr Andreas Zeitler)

Chapter 7

United Kingdom

ZH101 was the first E-3D delivered to the Royal Air Force. (Chris Lofting)

The UK's decision to order the E-3D was based on the final cancellation of the Nimrod AEW.Mk 3, and the issue of a new ASR issued in March 1986 for an AEW aircraft capable of tracking up to 400 surface and aerial contacts around 360° out to 275km (171 miles). The MoD initially ordered six E-3D Sentry AEW.Mk 1s, and later purchased a seventh example. The first example delivered was ZH101, which made its first flight on 5 January 1990, and the initial crews consisted of many officers and other ranks that had flown and trained with NATO's AEWF in 1986 and 1987.

A typical E-3D crew consists of some 18 members, depending on the mission, including the flight deck crew (pilot, co-pilot, navigator, and engineer), the mission crew (tactical director, fighter allocator, three weapons controllers, data-link manager, ESM manager, comms manager, comms operator, surveillance controllers, two surveillance operators, a radar technician, comms technician and a display technician). In an attempt to save money, instead of the three banks of consoles containing six operator positions as installed in the USAF and NATO E-3s, the RAF and the French Air Force only specified two banks of consoles in their aircraft.

Between 1998 and 2000 the E-3Ds were put through the Radar System Improvement Program (RSIP) at the cost of some GBP120 million. Improvements were made to the aircraft's systems in line with the USAF's Block 30/35 and at the same time new GPS navigation equipment was also installed. Then in January 2006, Boeing and Lockheed Martin were awarded contracts to conduct technology demonstrations to upgrade the mission systems on the seven E-3Ds, under Project Eagle. Various upgrades and long-term civil maintenance programmes have indeed been ongoing, the latest announced in May 2014 was the completion by Northrop Grumman of an upgrade to the E-3D fleet's IFF Mode S system that was included as part of the Whole Life Support Programme (WLSP) that the civil contractors have been scheduled to run until 2025. As of 2014, the RAF fleet is believed to include only five serviceable aircraft.

Sea King ASaC.Mk 7

As described in Chapter 4, the Royal Navy's Sea King AEW.Mk 2 system was planned, put together and test flown in a remarkably short period time in early summer 1982 and fulfilled an immediate need to provide an organic AEW capability for the fleet. The first two helicopters were joined by a further six that were converted to an established production standard that was eventually arrived at with the first two prototypes after extended trials and testing, some of which was carried out on the voyage to the South Atlantic. 849 NAS was officially reformed on 1 November 1984 at RNAS Culdrose, Cornwall. Over time it became apparent that although the Thorn-EMI ARI 5930/3 Searchwater radar was functional it did had severe limitations: the console displays were black and white monitors and the radar suffered badly from a ring of sea clutter in a circle below the aircraft. A major upgrade arrived in 2002 with the Cerberus system. The new system's main sensor is the Thales Searchwater 2000AEW, which was approved for use in the Sea King in October 1997.

The Searchwater 2000AEW was derived from the Searchwater 2000MR radar that had been designed for the stillborn BAE Systems Nimrod MRA.Mk 4 maritime patrol aircraft. The 2000AEW features a new emitter/transmitter set and a much lower weight, which increases mission endurance of the helicopter by an additional 45 minutes. It also has a much greater resistance to jamming and is able to simultaneously detect up to 300 air, ground and naval targets in three scans, each of which is dedicated to one of those individual types of target. The aircraft was also upgraded with the JTIDS/Link 16 system, which allows the radar operator to designate targets remotely to friendly fighters without any voice transmissions being required. The infamous clutter ring has now gone and the radar picture is displayed clearly on 53cm (21in) high-definition colour monitors.

The system is still affected by temperature inversion layers and coastal terrain pick-up but suppression and clever positioning can minimise their effects. During the mission planning process all information appertaining to the sortie is entered in the air-

Sea King ASaC.Mk 7 ZE420 conducts a confined-area landing on a cliff top in Cornwall in February 2012. (Nick Martin)

craft navigation system through the Whiskey Initialisation Support Peripheral (WISP). The WISP consists of a notebook computer, removable hard drives and diskette/CD readers. Data is automatically matched to the performance characteristics of the helicopter and enables detailed mission planning and debriefing. All of these upgrades determined a type designation change to the Sea King Air Surveillance and Control (ASaC) Mk 7.

The weight of all the new equipment and additional electronics pallets behind the two radar operators means that the crew normally consists of only one pilot and two observers.

Project Crowsnest

The Sea King ASaC.Mk 7 is due to be retired in 2016, after 10 years of service. On 5 September 2012 the UK Parliament Defence Committee announced that Project Crowsnest would be introduced to satisfy the requirement for an assured ASaC capability to provide long-range surveillance and battlespace management to carrier strike and littoral manoeuvre task groups. The selected mission system will be installed into existing Merlin HM.Mk 2 aircraft. Among the technical requirements for the system are the rapid installation/removal of radar and consoles on to and off the Merlin as required. There are four main competitors for the tender: Lockheed Martin (also the project manager, with its Vigilance radar), Thales UK (with an improved version of its Searchwater 2000 radar and Cerberus system), IAI Elta, and Selex ES. The proposals of the latter two contractors are currently under development. It is planned that eight Crowsnest packages will be acquired for installation in the Merlin fleet, replacing the 13 Sea King ASaC Mk. 7 systems currently in service.

Radar consoles inside Sea King ASaC.Mk 7 XV697. (Ian Shaw)

United States

The initial variant of the Sentry, the E-3A, was actually composed of two subtypes with different avionics mission suites: the Block 10 comprised 23 aircraft and was characterised by the AN/APY-1 radar and CC-1 computer, while the single example of the Block 15 had the AN/APY-2 radar with an expanded capacity against naval targets and the CC-2 computer with more memory and processing power. The next version was the Block 20 upgrade, which was introduced after the Sentry had been in USAF service for three years. The 24 original E-3A models received the new TADIL-C datalink, the Class 1 JTIDS and the Have Quick encrypted radio system. These modifications required that the mission crew be expanded to 17. Active self-defence in the form of chaff and flare dispensers were also fitted as were pylons for AIM-9 Sidewinder air-to-air missiles. These upgrades were sufficient for the modified airframes to be recategorised as E-3Bs, the first of which were delivered in mid-1984.

The Block 30/35 modifications introduced the AN/AYR ESM, 1 2H JTIDS terminals, CC-2E mission computer and a second GPS NAVSTAR system. This upgrade was completed in 2001 and covered the entire USAF E-3 fleet. The Block 30/35, or Radar System Improvement Program (RSIP), upgrade was evaluated in the Green Flag air exercise in 1999. One adverse feature of the new update was the decrease in the performance of the 'over the horizon' radar-operating mode, which was caused by pulse-Doppler compression.

In 2003, the USAF decided to promote a major update of its E-3C fleet, in order to achieve the following objectives: introduction of sensor/data fusion capability; diminish the intervals needed to engage time-sensitive targets; make it possible to carry out cheaper upgrades and to improve mission capability rate. This resulted in modification in five areas: sensors, IFF interface and ESM processor. With regard to sensors, a New-Generation Identification Friend or Foe, which is capable of Mode 5 interrogation and extends the effective range of the AWACS interrogator, while helping discriminate against closely spaced cooperative targets, is supplemented by an increase in the ESM's processing power and database. As for communication/navigation systems, the DRAGON (Replacement of Avionics for Global Operations and Navigation') was the main complex installed and was made up of a flight management system and joint mission planning system. These devices provide the E-3 with the required navigation performance, and the surveillance and communication capabilities necessary to maintain continued critical unrestricted access to global airspace into the future.

The modernisation of the mission computer encompassed the capability to automatically fuse all on-board and off-board sensor inputs to provide a single track for each air, sea, and land entity using a multi-sensor integration algorithm. It uses the 'one target, one track' philosophy, which merges all available on-board sensor data, furnishes an accurate air track picture and recommends rule-based identification to the operator. Among the sensors processed are the radar, IFF, ESM and Links 11/16, as well as operator inputs and air tasking orders. This upgrade also provides an update to the E-3 Link 16 and satellite communications capabilities, software to automatically refresh the on-board database, a renewed mission system health monitoring tool, better interfaces and controls of the on-board passive ESM system and mission planning and post-mission processing capabilities. The gathered data is presented on 15 monitors running the Windows NT operating system belonging to the Raytheon Solipsys

Operators on board an E-3B from the USAF's 965th Expeditionary Air Control Squadron track simulated hostile aircraft during an exercise in Southeast Asia in November 2008.
(SSgt Aaron Allmon/USAF)

USAF E-3C serial number 82-007 seen at March Air Reserve Base, California in 2007. Re-designated in 2011, the aircraft became the first production E-3G Block 40/45 Sentry.
(Nathan Havercroft)

Tactical Display Framework baseline, and which are distributed as follows, from the forward section: a block of five, another block of four and a final block of six monitors. The Data Link Infrastructure (DLI) is a software program that provides priority processing to reduce message transmission latency by transmitting high-priority messages first, taking the room previously occupied by a low-priority message in the queue.

A period of flight testing was performed by development aircraft TS-3 (E-3 serial number 73-1674) between April 2007 and July 2008 and included the installation of a large part of the new modifications aboard two aircraft of the 963rd AACS at Tinker.

During 2011, Boeing conducted developmental test and evaluation using the first production Block 40/45 E-3 (serial number 82-007). This consisted of seven flights from Boeing Field, Washington. It was found that Block 40/45 was operationally effective; it provided some improvements for the operators, but not all the required enhancements. Block 40/45 provided automated tracking and combat identification, but did not provide the operators with adequate control of the automated tracking capability. The crews were able to accomplish their battle management command and control missions throughout the evaluation. In addition, problems relating to the interoperability of the Link 16, mission computing and satellite communication were also reported. However, as was the case with previous modifications and upgrades applied to the E-3, the USAF was confident enough that it could resolve the outstanding issues and on 17 May 2011, 82-007 was officially re-inducted to service with the 552nd ACW, with the new service designation E-3G.

On 28 July 2014 the USAF announced the initial operational capability for the E-3G with the 552nd ACW. At the time, six Block 40/45 modified aircraft had been delivered to the wing and two E-3Gs (serial numbers 73-1674 and 82-007) had successfully deployed in support of counter-narcotics operations. All 31 remaining aircraft are scheduled to be upgraded by 2020, with the type's retirement date planned for 2035.

E-2C Hawkeye 2000

In December 1994 it was proposed to further improve the E-2C's operational capability with a mission computer upgrade, which was to replace the Raytheon Model 940 with four DEC 2100 Model A500MP systems. In general terms, the new electronics suite took up less than half of the space previously taken up by the L-304, and was three times lighter with a processing speed 15 times as fast. In the new system the operational flight profiles are loaded into the platform's navigational system via a mission information software transfer cartridge. This cartridge contains the navigation data, communication codes and relevant intelligence files for the planned mission. The new system was evaluated between January and July 1997, after having been installed aboard the second example of the E-2C Group II (BuNo 164109). The evaluations were not a complete success, with some failures occurring. The programme entered the evaluation phase aboard seven aircraft between 1997 and 2001 (six equipped the new mission computers, while the seventh had SATCOM, Cooperative Engagement Capability (CEC) and a new cooling system. The trials involved operations from land bases and aircraft carriers and in concert with the CEC-equipped cruiser USS *Anzio* (CG 68) which revealed various teething problems with the newly installed equipment.

This resulted in a further five Group II examples being added to the test programme between 1999 and 2001. Other modifications were also evaluated aboard E-2C BuNo 163849. They included satellite data transmission and communication equipment, the new cooling system and CEC. The latter resulted in the addition of a 1.37m (4ft 6in) antenna being installed underneath the Hawkeye's fuselage. The CEC antenna contains an omnidirectional transceiver that connects with command centres on parent aircraft carriers and surface combatants, thus providing an improved situational awareness or commanders and a uniform tactical picture for allied platforms. The CEC also provides accurate targeting data for ship-launched surface-to-air missiles.

The first Hawkeye 2000s were delivered in October 2001 and saw combat duty as part of VAW-117 aboard the USS *Carl Vinson* (CVN 70) over Afghanistan in 2002. As the problems were ironed out, 24 examples of this new version were ordered (21 for the US Navy, two E-2Ts for Taiwan, and one for France). Although designated Hawkeye

An E-2C Hawkeye 2000 from VAW-126 returns to USS *Harry S. Truman* (CVN 75) after a mission. The carrier was deployed in the Arabian Sea, close to Pakistan and Iran. (Georg Mader)

2000 or E-2C Group III at first, these aircraft are now officially known in the US Navy as E-2C Group 2 Plus and identified by the symbol '+' painted in white on the black nose of the aircraft. The original designation has now been reserved for foreign operators only. It should be noted, however, that none of the foreign Hawkeye 2000s have the CEC or SATCOM installations.

E-2D Advanced Hawkeye

The terrorist attacks of 11 September 2001 were catalyst for yet another modernisation of the E-2. Known as the Radar Modernization Program, it was contracted three months later with Northrop Grumman Integrated Systems, Lockheed Martin and Raytheon. The improvements introduced allow management of 'dynamic attacks' when the target has been changed at short notice, as well as an enhanced mission planning facility. These requirements resulted in a radical transformation of the Hawkeye's electronics suite, whose main component became the new Lockheed Martin AN/APY-9 AESA radar, housed in a rotating radome capable of 4.5 or 6 rotations per minute in order to hamper hostile interference. It uses a signal-processing technique known as STAP (Space-Time Adaptive Processing), meaning the adaptive nullifying of ground clutter and/or jamming to improve target detection in the presence of interference. The new device detects cruise missiles at longer ranges than its predecessor, offering 300 per cent greater search volume than the E-2C. It automatically distinguishes not only contacts in the sea and air, but also vehicles on the ground, a capability hitherto unheard of. The system was evaluated for the first time in July 2002 aboard an ex-US Coast Guard NC-130H, and the construction of the first five systems was initiated in the spring of the following year, having already demonstrated their resilience in an environment of strong clutter over land.

The first fully equipped example of the new aircraft, AA-1, was dedicated to the development and demonstration of the system and left the assembly line at St Augustine on 30 April 2007. Its maiden flight took place on 3 August 2007. While AA-1 is dedicated to structural testing, the second example, AA-2, was devoted to assessment of the mission equipment. The aircraft features a fully digital cockpit with three 43cm

An E-2D Hawkeye assigned to Test and Evaluation Squadron 1 (VX-1) lands aboard the USS *Dwight D. Eisenhower* (CVN 69). (Mass Communication Specialist Seaman Albert Jones/US Navy)

Seen during a test flight in October 2002, this NC-130H was serving as a test bed for the Radar Modernization Program (RMP) under development for the E-2D Advanced Hawkeye. (US Navy)

(17in) LCD screens, two of which are connected to special boards on the legs of the pilots, allowing any one of them have access to the same data as available to the mission crew, which consists of the Combat Information Center Officer, in charge of the mission command and planning, the Air Control Officer, in charge of supervising aircraft in flight, data-link and satellite communications, and the Radar Operator, who also manipulates other mission equipment and assists the Air Control Officer, all have with 51cm (20in) active-matrix LCD monitors. The previously described CEC capacity was also introduced, with all equipment connected via Mil Std 1553 digital bus and optical fibre intranet.

In-flight refuelling is planned to be added to the E-2D in Fiscal Year 2019. This will employ Northrop Grumman's Advanced Hawkeye Aerial Refueling system that will equip new-production E-2Ds and be retrofitted to E-2Ds previously delivered. Several system upgrades are associated with the new aerial refuelling capability, including new pilot seats, formation lights and enhanced flight-control software.

The first example of the 'Hawkeye Delta' was delivered to VAW-120 at NAS Norfolk on 29 July 2010. A total of 75 of these aircraft have been procured by the US Navy to serve in five-aircraft detachments aboard its aircraft carriers. VAW-125 is the first operational unit equipped with the E-2D, declaring initial operational capability in autumn 2014.

An E-2D Advanced Hawkeye launches using the electromagnetic aircraft launch system (EMALS) at the full-size shipboard-representative test site at Joint Base McGuire-Dix-Lakehurst in New Jersey in September 2011. (Kelly Schindler/US Navy)

APPENDIX I

Technical specifications of AEW aircraft

The very nature of AEW&C aircraft and their role dictates that the countries that operate them keep the accurate data regarding their aircraft's capabilities and vulnerabilities concealed from potential enemies. Therefore the arts of disinformation and secrecy are frequently used when it comes to published data that becomes available regarding these platforms especially ones that are currently in service around the world. Through this mine field any researcher must be very wary of any touted or boasted of enhanced performance of radar or airframes. With this in mind this appendix can be best described as an amalgam of reliable known and published data of the individual aircraft and a fair judgement of what it may or may not be capable of achieving. Many sources were used in its compilation but where available the majority of data was obtained from several volumes of Jane's All the World's Aircraft. Where data is not available or is genuinely suspect, it has been omitted.

AgustaWestland AW101 Mk 112

Powerplant
Three General Electric T700-GE-T6A turboshafts each rated at 1,394kW (1,870shp)

Dimensions
Length, fuselage 19.53m (64ft 1in); width 4.6m (15ft 1½in); height, both rotors turning 6.62m (21ft 8¾in); main rotor area 271.51m² (2,922.5 sq ft)

Weights
Empty 10,500kg (23,148lb); maximum 15,600kg (34,391lb)

Performance
Maximum speed 309km/h (191mph); cruising speed 278km/h (173mph); rate of climb 852m (2,795ft) per minute; service ceiling 4,575m (15,000ft); range 1,390km (863 miles)

Avionics
Smith Industries Alenia OMI SEP 20 dual redundant AFCS, BAe Systems LINS 300 ring laser INS, MARA aircraft management computer, three Marconi Selenia Communications SRT-651 V/UHF and two SRT-170L HF radios, MSC Link 11 TD8503 datalink terminal, MSC SP-1450 intercom, Italtel Mk 12 IFF/transponder, Elettronica ALR-735(3) electronic support system, BAe Systems RALM/1 laser warner, Elettronica ELT/156X(V2) RWR, MSC Doppler, MSC R-202 GPS, Litton Italia LISA 4000D strapdown attitude/heading reference system, Galiflir FLIR

Radar
FIAR Eliradar HEW-784 360° Doppler air and surface surveillance radar (based on MM/APS-784), operating in X-band (8-12.5GHz), installed inside 3m (9ft 10in) radome mounted under forward fuselage. Designed to detect and track up to 128 aerial targets (with look-up/look-down capability), 64 land/sea targets, presented on two Alenia/Racal screens

Modes: surveillance look-down, detection/tracking of high radial speed targets in presence of ground clutter; surveillance look-up (clear); ground mapping at short/medium/long range; alarm (missile); detection and tracking of sea-surface. Radar range 200km (123 miles)

Production
4

Crew
Four, consisting of two flight crew and two radar operators

Operator
Italy

Avro Shackleton AEW.Mk 2

Powerplant
Four 12-cylinder Rolls-Royce Griffon Mk 58 piston engines, each developing 1,462kW (1,960hp) or 1,816kW (2,435hp) with water-methanol injection, driving eight contra-rotating de Havilland propellers each of 3.96m 13ft) diameter

Dimensions
Wingspan 36.53m (119ft 10in); length 26.62m (87ft 4in); height 5.10m (16ft 9in)

Weight
Empty 26,218kg (57,800lb); maximum take-off 45,360kg (100,000lb)

Performance
Maximum speed 333km/h (207mph); cruising speed 296km/h (184mph); service ceiling 6,200m (20,340ft); maximum range 5,000km (3,100 miles)

Avionics
A1961 intercom, PTR 175 V/UHF radios, two 618T HF radios, AN/APX-7 IFF, AN/ARC-52 UHF radio, Mk 7B radio altimeter, ILS, AD712 ADF, TACAN, LORAN, Decca Doppler 72, ARI 18144 ESM/Orange Harvest (with two 16613 CRT-type monitors), P12 magnetic compass, Mk 2A accelerometer, ADD stall alert, ADRIS device, two AN/APA-57B and one AN/APA-5A 17.8cm (7in) monitors

Radar
AN/APS-20F(I) with AMTI, S-band (2,880MHz), NATO Band E, performance similar to that fitted to Fairey Gannet AEW.Mk 3, peak power 1.75MW, capable of detecting destroyer-type targets at up to 370.6km (230 miles), fighter-type targets at 120.5km (75 miles), medium bomber at 157.5km (98 miles)

Appendix I

Production
12

Crew
Nine, consisting of two pilots, one air engineer, two navigators (one manning the radios), and four mission crew comprising of tactical coordinator (TACO) controller and three radar operators

Operator
United Kingdom

Beriev A-50, Baghdad/Adnan and Xi'an KJ-2000

Powerplant
(All variants, except A-50EI) four Aviadvigatel D-30KP Series 1 or 2 turbofans each rated at 12,000-12,500kg (26,465-27,557lb) of thrust at take-off; A-50EI four Aviadvigatel PS-90A-76 turbofans each rated at 14,500kg (31,967lb) of thrust

Dimensions
Wingspan (all variants except KJ-2000) 50.5m (165ft 6in); length, all variants (without in-flight refuelling probe) 46.59m (152ft 10¼in), (with in-flight refuelling probe) 49.59m (162ft 8in); height (all variants except A-50EI) 14.76m (48ft 5in), A-50EI 14.80m (48ft 7in)

Weights
(Estimated), empty (all variants) 75,000kg (165,347lb); maximum take-off (all variants except KJ-2000) 170,000kg (374,786lb); KJ-2000 195,000kg (429,900lb)

Performance
Maximum speed 800km/h (497mph); patrol speed 600km/h (372mph); service ceiling 10,200m (33,465ft); ferry range 7,714km (4,793 miles), patrol range (all variants except KJ-2000) 5,745km (3,569 miles); KJ-2000 5,000km (3,107 miles)

Avionics
A-50 and A-50U: 1T SALT-2BT automatic flight control system, DISS-013-S2M or DISS-013S2 Doppler sensor, RLS-N weather radar, RLS-P terrain mapping radar, TKS-P precision compass, P-76 Double INS, RSBN-7S short-range navigation system, Koors MP-70 and Koors-MP-2 automatic approach systems, five RV/RV-5M radio altimeters, ARK-15M automatic direction-finder, SDK-67/SD-75 distance meters, R-855A1/R-855UM/R-861 UHF radios, Mikron VHF radios, SPU-8/SPU-15 flag radio, intercom, RI-65 oral warning system, Mars-BM and MS-61B voice flight data recorder, S-3M Sirena-2 electronic support measures, SRO-1P, SO-70, SO-72M and SOUND-64 IFF transponders

Baghdad 1 and 2, Adnan 1 and 2 and Simorgh: avionics equipment not disclosed but believed to be of Italian (Selenia) and British (Marconi) origin; Collins IFF

A-50EI and KJ-2000: no details of mission avionics

Radar
A-50: originally equipped with Liana radar as used in Tu-126, with similar performance. 360° scan and simultaneously detection of 50-60 contacts and control of 10 concurrent intercepts via data-link

A-50U: upgraded with new Vega Shmel-2 radar. Three-dimensional pulse-Doppler radar, 4GHz frequency, 360° scan. Equipped with passive mode to detect ECM sources without transmission-induced vulnerability, digital MTI function that indicates all fighter-sized airborne targets out to a search radius of 230km (143 miles) or 400km (248 miles) for large aircraft and surface ships; control of 15 concurrent intercepts via data-link

Baghdad 1 and 2, Adnan 1 and 2 and Simorgh: licence-built Thomson-CSF TRS-2105/06 Tigre-G, two-dimensional radar operating in G/H- band, claimed to be able to detect 65 fighter-sized contacts at 100km (62 miles) and with a maximum range of 350km (217 miles)

KJ-2000: AESA radar, reportedly designated Type 88, designed by Nanjing Research Institute of Electronic Technology (NRIET), which is also known as No. 14 Institute. Radar employs three antennas mounted in fixed radome, comprising a three-dimensional electronically steered array with 360° coverage, operating in frequencies between 1,200-1,400MHz with maximum target detection range of 470km (292 miles). Radar reportedly based on experience gained from No. 14 Institute's earlier indigenously developed Type H/LJG-346 shipborne active phased-array radar system (SAPARS) that was completed in 1998. This in turn was the predecessor of the AESA system that equips the PLAN's Lanzhou-class destroyers.

A-50EI: AHMED EL-2075 Phalcon operating in D band (1 to 2GHz), variable frequency, electronic tracking in elevation and azimuth with the following operating modes, which can be interleaved: passive, high-frequency search, TWS, low-frequency naval search and slow screening (for helicopters); can detect 100 contacts simultaneously; range of 400km (249 miles) for aircraft and ships and 80km (112 miles) for helicopters

Production
25 A-50, 3 A-50EI, 4 Baghdad/Adnan, 4 KJ-2000

Crew
Five flight crew, plus mission crew of 10 to 15

Operators
China, India, Iran, Iraq and Russia

Boeing 737 AEW&C

Powerplant
Two CFM International CFM56-7B27A turbofans each rated at 12,383kg (27,000lb) thrust

Dimensions
Wingspan 35.78m (117ft 2in); length 33.62m (110ft 4in); height 12.54m (41ft 2in); wing area 91.45m² (980 sq ft)

Weights
E-7 Wedgetail: Empty 46,607kg (112,750lb); maximum take-off 77,565kg (171,000lb); other operators not known

Performance
Maximum speed 870km/h (540mph); cruising speed 760km/h (473mph); service ceiling 12,496m (41,000ft); range in excess of 6,482km (4,030 miles) without in-flight refuelling

Avionics
E-7 Wedgetail: six Honeywell flat-panel liquid crystal flight displays, Smiths Industries dual flight management systems, Rockwell Collins Series 90 avionics suite with dual ADF, TCAS II, Rockwell Collins ILS/VOR/DME, Raytheon GPS, three BAE Australia/Elta Systems ALR-2001 ESMs, Northrop Grumman AN/AAQ-24(V) Nemesis DIRCM, four Northrop Grumman AN/AAR-54 MAWS, four BAe Systems AN/ALE-47 ACDS, four Elisra LWS-20 laser warning systems, three Rockwell Collins AN/ARC-220 HF voice/data channels, 10 Rockwell Collins VHF/UHF (Have Quick II) AN/ARC-210 radio channels, four Rockwell Collins AN/URC-138 UHF terminals compatible with MILSATCOM voice/data, Link 11, OTCIXS and JTIDS/Link 16, all managed by AvTech Tyee 5346 pan-

els, Esterline 2100 secure intercom, five consoles with two operator stations, each with a 53cm (21in) flat-panel monitor, one 15.9cm (6.25in) square tactical monitor on flight deck, two redundant Sun Microsystems Blade mission computers

Korea: similar to E-7 Wedgetail, except for electronic mission equipment of Korean origin

Turkey: similar to E-7 Wedgetail, except for the following: Havelser mission computer, three Selex Galileo SRT-470 HF radios, 10 Aselsan software-defined V/UHF radios, Avelsan/Mikes ESM, Aselsan ASES self-defence system (comprising missile warning system, RWR and countermeasures dispenser system)

Radar
Northrop Grumman MESA D-band (1.2-1.4GHz) radar with 360° coverage, capable of detecting up to 3,000 air/surface/sea targets simultaneously within a range of 350km (217 miles); range doubled when dedicated sector mode is activated. Other modes include: spot, low PRF, high PRF, uniform coverage, ground stabilised sector, platform stabilised sector, track revisit. The MESA is capable of simultaneously tracking 180 targets and conducting 24 intercepts and incorporates an IFF with a range in excess of 555km (345 miles)

Production
14 (plus 3 on order)

Crew
12+, consisting of two flight crew and eight or more mission crew.

Operators
Australia, Qatar (on order), South Korea and Turkey

Beyond the Horizon – The History of Airborne Early Warning

Boeing E-3 Sentry

Powerplant
E-3A/B/C/G four Pratt & Whitney TF33-PW-100A turbofans each rated at 9,525kg (21,000lb) thrust; E-3D/F and RSAF E-3A four CFM International CFM56-2A3I turbofans each rated at 10,886kg (24,000lb) thrust

Dimensions
Wingspan (E-3A/B/C/F/G) 44.45m (145ft 9in), E-3D 44.98m (147ft); length (all models) 46.61m (152ft 11in); height E-3D/F 12.60m (41ft 4in), E-3A 12.7m (41ft 9in), E-3B/C/G and RSAF E-3A 13.00m (42ft 6in); wing area (all models) 283.4m² (3,050 sq ft)

Weights
Empty, E-3B/C/G and NATO E-3A, 73,480kg (161,992lb), E-3D/F and RSAF E-3A 83,990kg (185,163lb); maximum take-off E-3B/C/G, 156,150kg (344,246lb), NATO E-3A 147,429kg (325,000lb), RSAF E-3A 156,000kg (343,915lb), E-3D 150,822kg (332,500lb), E-3F 151,950kg (334,986lb)

Performance
Maximum speed, E-3B/C/G 805km/h (499mph), NATO E-3A and RSAF E-3A 827km/h (513mph), E-3D 851km/h (528mph), E-3F 854km/h (530mph); cruising speed E-3E/B/C/G and NATO E-3A 579km/h (359mph), E-3D/F and RSAF E-3A 763km/h (472mph); service ceiling (all models) in excess of 8,839m (28,999ft); maximum range (all models) in excess of 9,251km (5,748 miles)

Avionics
E-3A: two AN/ARC-165 HF radios, three AN/ARC-173 VHF radios, two AN/ARC-171 UHF radios, one AN/ARC-181 crypto radio (for JTIDS/TADIL-A/Link 11 system, later replaced by URC-107(V)4 radio, standard 2H, also installed on USAF and foreign updated models), OMEGA AN/ARN-120 navigator, AN/APN-213 Doppler navigator, AN/ASN-118 navigation computer, AN/ASN-119 INS, 51RV-2B/3418 VOR/ILS, AN/AYC-1 digital data processor, AN/AIC-28(V)1 intercom (updated variants introduced in subsequent models), CC-2 mission computer, AN/APX-103 IFF interrogator, nine Hazeltine workstations (with 48cm/19in CRT screens).

RSAF E-3A: similarly equipped, but with Link 16, and URC-107(V) 4 radios operating under 2H standard

E-3B: similar to E-3A apart from three AN/ARC-230 HF radios (Link 11 standard), three AN/ARC-166 VHF radios, eight AN/ARC-169B UHF radios, 14 workstations similar to those aboard E-3A

E-3C: similar to E-3A/B but with two AN/ARC-230 HF radios (Link 11 standard), three AN/ARC-168 VHF radios (later replaced by AN/ARC-210), AN/ARC-204 radio (V model, Have Quick II standard/A-NET network), 14 ELCAN/Raytheon workstations with 53cm (21in) CLADS LCD screens, CC-2E mission computer, GPS (NAVSTAR at first, but soon replaced by GPS/INS 'GINS', also aboard foreign updated variants), AN/APX-103B IFF, AN/AYR-1 device for SIGINT/ELINT

E-3D: similar to E-3A, but with three AN/ARC-2201 HF radios, AN/ARC-204 radio (as in E-3C), JTIDS/IJMS systems compatible with Links 4, 11, 14 and 16, MDES operational complex, EW-1017 Yellowgate ESM, nine Racal consoles with 51cm (20in) screens

E-3F: similar to E-3A, but with AEG/Siemens workstations, 51cm (20in) screens, three AN/ARC-166 VHF radios, AN/AYR-1 ESM, JTIDS/IJMS/Link 16-compatible communications

NATO E-3A: similar to E-3A, but with AN/ASC-40 suite (five RT-651/N-V V/UHF radios, for SATCOM and Link 16-compatible communications), Have Quick II crypto communications (with four AN/ARC-164 UHF radios), AN/AAQ-24 LAIRCM, NATO E-3A IFF, STR 2000/TSC 2000 IFF/transponder, computing devices, workstations with 51cm (20.1in) screens

Radar
Westinghouse AN/APY-1/2 radar (E/F-band) operating in the following modes: (1) pulse-Doppler non-elevation scan able to detect targets at a range of 394km (245 miles), (2) pulse-Doppler elevation scan, (3) beyond-the-horizon, (4) passive, (5) standby, (6) maritime. Modes 2 and 3 can be interleaved. Management of up to 60 aerial interceptions and simultaneous detection of 600 targets, with radar range exceeding 400km (249 miles).

Production
67

Crew
21+, consisting of four flight crew and 17+ mission crew

Operators
France, NATO, Saudi Arabia, United Kingdom and United States

Boeing E-767

Powerplant
Two General Electric CF6-80C2B6FA turbofans, each rated at 27,900kg (61,500lb) thrust

Dimensions
Wingspan 47.57m (156ft 1in); length 48.51m (159ft 2in); height 5.85m (52ft); wing area 283.3m² (3,050 sq ft)

Weights
Empty 87,000kg (191,198lb); maximum take-off 174,635kg (385,000lb)

Performance
Maximum speed 805km/h (500mph); cruising speed 722km/h (449mph); service ceiling 12,220m (40,100ft); range 9,290km (5,754 miles)

Avionics
Two Litton LN-100G INS/GPS platforms, Honeywell navigation suite including RDR-4A weather radar, EFIS-700 electronic flight instruments system, FCS-700 flight control system, Hazeltine DME, AN/ASN-168 GPS/INS, 14 BAE Systems workstations, AN/ARC-207 HF radios, AN/ARC-243V transceiver and AN/ARC-228 receptor HF radios (Link 4 and 11/TADIL-C/JTIDS), AN/ARC-171 UHF encrypted radios (Have Quick), AN/ARC-166 VHF radios, AN/ARC-227 SATCOM equipment, AN/AIC-28(V)X intercom, ARGO Systems AN/AYR-1 SIGINT/ELINT sensor, IBM CC-2E central mission computer, Eaton (AIL) AN/APX-103 interrogator set with airborne IFF interrogator, Mk X SIF air traffic control and Mk XII military IFF in a single integrated system. Interrogator, capable of detecting contacts located at up 800km (497 miles) mounted in the same rotodome as the radar

Radar
Northrop Grumman AN/APY-2 pulse-Doppler radar mounted in a 7.32m (24ft) span, 1.52m (5ft) deep rotordome above the fuselage operates in the E/F (2-4GHz) bands in the following modes: (1) pulse-Doppler non-elevation scan able to detect targets located at 394km (245 miles), (2) pulse-Doppler elevation scan, (3) beyond-the-horizon, (4) passive, (5) standby, (6) maritime. Modes 2 and 3 can be interleaved. The E-767 can manage up to 60 aerial interceptions and simultaneously detect 600 targets, with radar range exceeding 400km (249 miles)

Production
4

Crew
21+ consisting of pilot, co-pilot and up to 19+ mission crew

Operator
Japan

Boeing PB-1W

Powerplant
Four Wright R1820-97 Cyclone turbo-supercharged radial engines, each rated at 895kW (1,200hp)

Dimensions
Wingspan 31.60m (10 ft in); length 22.60m (74ft 4in); height 5.82m (19ft 1in); wing area 131.92m² (1,420 sq ft)

Weights
Empty 17,650kg (38,911lb); maximum take-off 28,865kg (63,636lb)

Performance
Maximum speed 475km/h (296mph); cruising speed (AEW operations) 411km/h (255mph); rate of climb 231m (760ft) per minute; service ceiling 9,754m (32,000ft); range 4,778km (2,969 miles) with auxiliary tanks

Avionics
AN/AIC-2 intercom, SCR-287 long-range radio, SCR-274-N short-range radio, SCR-535A or SCR-695 IFF (in later aircraft), RCM radar countermeasures, SCR-522 HF radio, RC-43 beacon, SCR-269-G radio compass, SCR-718/AN/APN-1 radio altimeter; AN/APN-9 LORAN receiver, RC-103 and AN/ARN-5 ILS, SCR-578-A emergency radio, AN/ART-28 radar transmitter, AN/APA-56, -57 and -81 screens

Radar
AN/APS-20A. Slightly improved compared to AN/APS-20, but with same overall features

Appendix I

Production
22

Crew
13: pilot, co-pilot, crew chief, navigator, CIC controller, three radar controllers, two radio operators, radar technician, mechanic, observer/passenger

Operator
United States

Changhe Z-18Y Black Bat*

* primary data refers to Aérospatiale SA321 Super Frelon, first acquired by China from France in the mid-1970s and subsequently built as the indigenous Z-8. The latest, much improved military version is the Z-18 series. The Z-18 itself is a derivative of the civil Avicopter AC313 first flown in March 2010.

Powerplant
Three turboshafts, reportedly upraded Changzhou WZ-6s, indigenous copies of the Turboméca Turmo 3C III. Basic WZ-6 rated at 1,130kW (1,536shp) for take-off

Dimensions
Length 23.03 m (75ft 6 5/8in); main rotor diameter 18.90m (62ft 0in); height 6.66m (21ft 10¼ in), main rotor disc area 280.6m^2 (3,019 sq ft)

Weights
Empty 6,863kg (15,130lb); maximum take-off 13,000kg (28,660lb)

Performance
Maximum speed 249km/h (155mph); rate of climb 400m (1,312ft) per minute, service ceiling 3,150m (10,325ft), range 1,020km (632 miles), endurance 4 hours

Avionics
Unknown.

Radar
Reported to utilise a surveillance radar system developed by the 38th Institute. Radar, thought to be a long-range multi-mode AESA type, is housed in a retractable installation located behind an aerodynamically shaped fairing on redesigned rear fuselage doors, similar to the system installed in the French AS532 Cougar Horizon battlefield surveillance helicopter

Production
At least 2 (4+ planned)

Crew
Five

Operator
China

Douglas AD-3W, 4W and 5W Skyraider

Powerplant
One Curtiss-Wright R-3350-26WA Cyclone radial engine rated at 2,013kW (2,700hp)

Dimensions
Wingspan 15.24m (50ft); length, AD-3W and AD-4W 11.70m (38ft 5in), AD-5W 12.19m (40ft); height, AD-3W and AD-4W 3.65m (12ft), AD-5W 4.77m (15ft 8in); wing area 37.19m² (400.31 sq ft)

Weights
Empty, AD-3W 5,896kg (13,000lb), AD-4W 5,715kg (12,600lb), AD-5W 5,484kg (12,092lb); maximum, overload, AD-3W 8,300kg (18,300lb), AD-4W 10,886kg (24,000lb), AD-5W 11,340kg (25,000lb)

Performance
Maximum speed, AD-3W 555km/h (345mph), AD-4W 564km/h (351mph), AD-5W 492km/h (306mph); cruising speed, AD-3W/AD-4W 253km/h 158mph), AD-5W 250km/h (155mph); service ceiling, AD-3W 9,396m (27,000ft), AD-4W 11,978m (36,000ft), AD-5W 8,168m (26,800ft); rate of climb, AD-3W/AD-4W 533m (1,748ft) per minute, AD-5W 841m (2,760ft) per minute; maximum range, AD-3W, AD-4W 2,315km (1,438 miles), AD-5W 3,329km (2,069 miles)

Avionics
AD-3W: AN/APX-1 IFF transponder (replaced by AN/APX-2 and later by AN/APX-6), AN/ARC-1 VHF receiver transmitter (replaced by AN/ARC-27 UHF radio in 1951), AN/ARC-5 command radio set, AN/APN-1 radar altimeter, AN/ARC-1 VHF receiver transmitter, ARR-2A navigation receiver (replaced by TACAN in 1955), AN/AIC-4 inter-

phone, AN/APX-13 IFF interrogator, AN/ART-26 radar relay, AN/ARC-28 UHF receiver transmitter, AN/APA-56 set comprised of AN/APA-57 ground position indicator and AN/APA-5A radar screen

AD-4W and Skyraider AEW.Mk 1: AN/ARC-1 VHF receiver transmitter, APN-2 radar altimeter, ARR-2A navigation receiver, AN/AIC-4 interphone, AN/ARC-5 command radio set, AN/APX-13 IFF, AN/ART-26 radar relay, AN/ARC-28 UHF receiver transmitter (from 1951 onwards)

AD-5W: AN/ARC-1 VHF receiver transmitter, AN/ARC-2 UHF transmitter receiver, AN/AIC-4 interphone, AN/ARN-6 LF ADF, AN/APX-6/7 IFF, AN/ARN-12 marker beacon, AN/APX-13 IFF, AN/APN-22 radio altimeter, AN/ARA-25 UHF ADF, AN/ARC-27A UHF transmitter receiver relay (from 1951 onwards), AN/ART-28 radar relay, AN/APA-57 ground position indicator, AN/APA-89 IFF coder

Radar
AD-3W, AD-4W and Skyraider AEW.Mk 1: AN/APS-20 radar, S-band (2-4GHz), 360° rotation, 6 or 10rpm selectable tracking velocity, with designed maximum range of 370km (230 miles) which was very rarely achieved and then only with very large targets

AD-5W: AN/APS-20B radar, with 2MW power source

Production
31 AD-3W, 168 AD-4W (including 50 for UK), 128 AD-5W

Crew
Onw pilot and two radar operators

Operators
United Kingdom and United States

Embraer 145 AEW&C

Powerplant
Two Allison 3007A1P turbofans providing a combined thrust of 7,564kg (16,675lb) on take-off

Dimensions
Wingspan 21.0m (68ft 11in); length 29.87m (98ft); height 7.22m (23ft 6in); wing area 52.00m² (550.9 sq ft)

Weights
E-99: empty 14,800kg (32,628lb); maximum take-off 23,400kg (51,587lb)
EMB-145H: empty 16,000kg (35,274lb); maximum take-off 23,400kg (51,587lb)

Performance
Maximum speed in excess of 834km/h (518mph); rate of climb in excess of 780m (2,560ft) per minute; service ceiling in excess of 11,278m (37,000ft); maximum range, E-99 2,664km (1,655 miles), EMB-145H 2,226km (1,383 miles); endurance 6 hours+

Avionics
E-99: Thales TSA/TSB 2500 IFF with a range of 450km (280 miles), Honeywell Primus 1000 avionics suite, Universal UNS-1D FMS, Honeywell H746G inertial/GPS navigation suite, L3 communications exploration system (C/NCES), DRS communication subsystem, five Rohde & Schwarz M3AR 400U XT452U3 VHF/UHF radios, three Ericsson workstations with Barco displays (one 49.5cm/19.5in MPRD 9651, and two 26.4cm/10.4in MPRT 126s installed on each), AN/AAQ-22 Star SAFIRE II FLIR, Calquest CQ 200 satellite data-link system

EMB-145H: Thales HF/V-UHF communication suite (compatible with Link 11, Link 14, and EuroMIDS LVT Link 16), ITT tactical terminal; Elettronica ALR-733(V)5 ESM/SIGINT, MBDA ELIPS-ADS countermeasures defence system, ISI Hellas data-link, five Ericsson workstations similar to those installed in E-99, Thales SF 25 Mk X IFF (interrogating in modes 1, 2, 3/A, 4 and 5), M3AR 400U XT452U3 VHF/UHF radio transmitters and receivers, BR2 data-link system, Greek version of Elettronica ALR/755(V)5 ESM, Link 11/14/16 data-links, Thales UHF radios, TSA/TSB 2500 combined interrogator/transponder, KY-100 communication set with KG-40 encryption modules

Radar
E-99: Ericsson PS-890 Erieye active phased-array electronic scan pulse-Doppler radar in the S-band (3.1 to 3.3GHz), AESA radar, with 300° area coverage. Operates in E and F bands. Radar can simultaneously detect 300 contacts, moving from below 108km/h (67mph) to over 3,600km/h (2,237mph), it can prioritise 100 targets and manage at least nine intercepting aircraft. Modes: extended early warning, track-while-scan, air-to-air, sea surveillance, frequency agility, target radar-cross-section display, ground stabilised or platform stabilised search. Radar has range of 150km (93 miles) against cruise missiles, 300km (186 miles) against surface vessels and over 450km (280 miles) against fighters.

EMB-145I: PS-890 Erieye capable of detecting up to 300 air, maritime and ground contacts at a maximum range of 400km (249 miles). Modes: search, track-while-scan, ground mapping, priority tracking and high performance tracking, with more than one mode being operated simultaneously; hovering helicopters detection capability. Secondary surveillance radar (Mk XII class), with a range of 520km (323 miles)

Production
13

Crew
E-99: (5) pilot, co-pilot, tactical coordinator, electronic warfare operator and intercept controller
EMB-145II: (7) pilot, co-pilot, signal intelligence specialist, radar operator, mission commander and two weapons controllers

Operators
Brazil (E-99), Greece (EMB-145H), India (EMB-145I) and Mexico (EMB-145AEW&C)

Fairey Gannet AEW.Mk 3

Powerplant
One Armstrong Siddeley ASMD.8 Double Mamba 112 turboprop, rated at 2,759kW (3,700hp); each unit rated at 1,380kW (1,850shp). Two contra-rotating Dowty Rotol propellers each of 3.82m (12ft 6in) diameter

Dimensions
Wingspan 16.60m (54ft 7in) spread, 6.0m (19ft 11in) folded; length 13.41m (44ft); height 5.1m (16ft 9in); wing area 44.87m² (483 sq ft)

Weights
Empty 6,395kg (14,098lb); maximum take-off 11,793kg (26,000lb)

Performance
Maximum speed 520km/h (320mph); single-engine cruising speed 233km/h (145mph); service ceiling 7,600m (25,000ft); range 1,296km (700 miles), endurance 5 hours flying at 233km/h (145mph) at an altitude of 914m (3,000ft)

Avionics
ARI 18124/1 UHF radio, ARI 18032 or ARI 23090 HF radio (after modernisation), ARI 5378 Mk 5 or ARI 18215/1 Mk 5A radio altimeter (after modernisation), ARI 18107 TACAN, ARI 5848 secondary radar (IFF) or AN/APX-7 (Mk 10, installed before modernisation), or ARI 23134/3 (after modernisation), ARI 5885 Doppler radar, AN/ART-28 data binding, two AN/APA-57B navigation devices, ADRIS device (airline unit, air position indicator Mk 1B, Mk 1c and 4c ground position indicators), Mk 1 wind broker unit, intercom system (with amplifier after 1961), ARI 5491 VHF receiver transmitters, ARI 5206 HF transceiver, ARI ZBX 5307 radio, AYF ARI 5284 low-altitude radio altimeter, Mk 4B magnetic compass with backup Type E2A (E2B) compass, two AN/APS-56 monitors

Radar
AN/APS-20F radar, later AN/APS-20F(I). S-band (2,880MHz), NATO Band E, peak power 1.75MW. Capable of detecting destroyer-type targets at distances up to 370.6km (230 miles), fighter-type targets at 120.5km (75 miles), medium bomber at 157.5km (98 miles)

Production
45

Crew
One pilot and two observer/ radar operators

Operator
United Kingdom

Goodyear ZPG-2W/3W

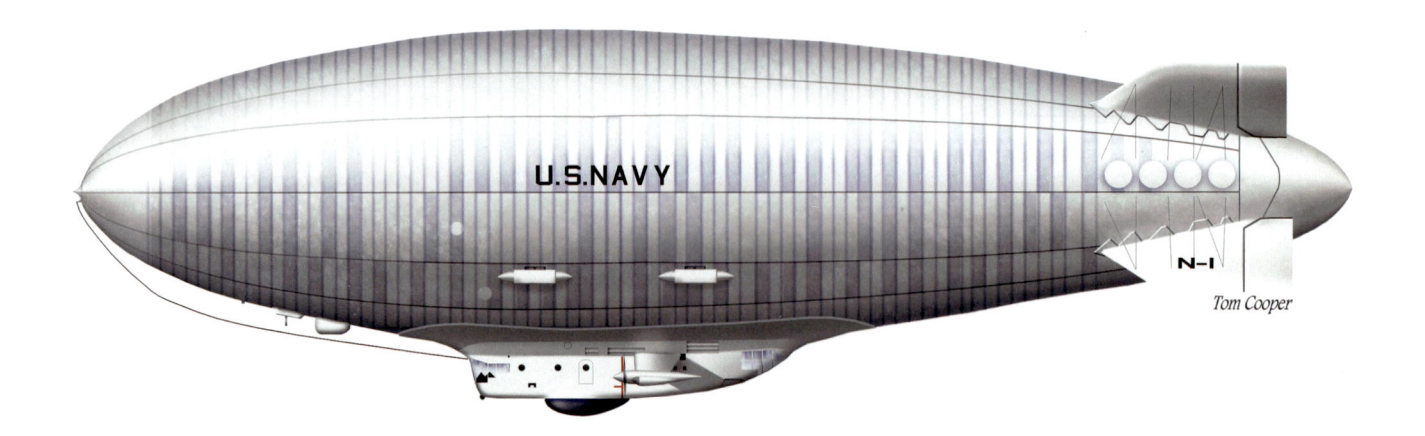

Tom Cooper

Powerplant
ZPG-2W: two Wright R-1300-2A radial engines each rated at 597kW (800hp)
ZPG-3W: two Wright R-1820-88 Cyclone radial engines each rated at 1,174kW (1,575hp)

Dimensions
ZPG-2W: diameter 22.98m (75ft 5in); length 104.41m (342ft 7in), height 32.61m (107ft 3in)
ZPG-3W: diameter 25.90m (85ft); length 117.65m (386ft); height 35.66m (117ft)

Weights
ZPG-2W: empty 21,630kg (47,700lb), maximum 29,302kg (64,600lb)
ZPG-3W: empty 32,241kg (71,080lb), maximum 42,230kg (100,000lb)

Performance
ZPG-2W: maximum speed 126km/h (79mph); cruising speed 74km/h (46mph); service ceiling 914m (3,000ft); range at optimum speed 3,574km (2,223 miles)
ZPG-3W: maximum speed 151km/h (94mph); cruising speed 83km/h (52mph); service ceiling 1,524m (5,000ft); rate of climb 731m (2,400ft) per minute; range at optimum speed 5,185km (3,222 miles)

Avionics
ZPG-2W: AN/APN-22 radio altimeter, AN/ARC-27A UHF transceiver, AN/APN-70 LORAN, AN/ARC-5 command radio set, AN/APX-5 and -7 IFF, AN/ARN-6 radio compass, AN/AIC-7A intercom, AN/APR-9B ECM receiver, AN/ALR-5 radar receiver, Loral AN/APA-74 ECM pulse analyser group, AN/APA-69A ECM direction-finding radar set, AN/ALA-2 ECM panoramic adaptor, AN/ARN-12 marker beacon, two AN/ART-13 transmitters, two AN/ARR-15 receivers, AN/APA-113 radar indicator system, AN/AVQ-12 searchlight

ZPG-3W: AN/ARR-41 HF receiver, AN/ARC-38 HF radio, CP-265 AN/ASA-14 dead-reckoning tracer, CP-262 AN/ASA-13 navigational computer group, C-1 compass, OS-4B/AP oscilloscope, AN/APN-70 LORAN, AN/APA-56 monitor set, AN/APS-69 height-finder radar with maximum altitude of 9,144m (30,000ft)

Radar
ZPG-2W: Lavoisier AN/APS-62 X-band radar, with range of 370km (230 miles), alternatively Hazeltine AN/APS-20B/E search radar, with same range as AN/APS-20

ZPG-3W: Hazeltine AN/APS-95 P-band 360° radar, with normal/improved detection modes and a range of 463km (288 miles)

Production
ZPG-2W (5), ZPG-3W (4)

Crew
21 to 25

Operator
United States

Grumman E-1 Tracer

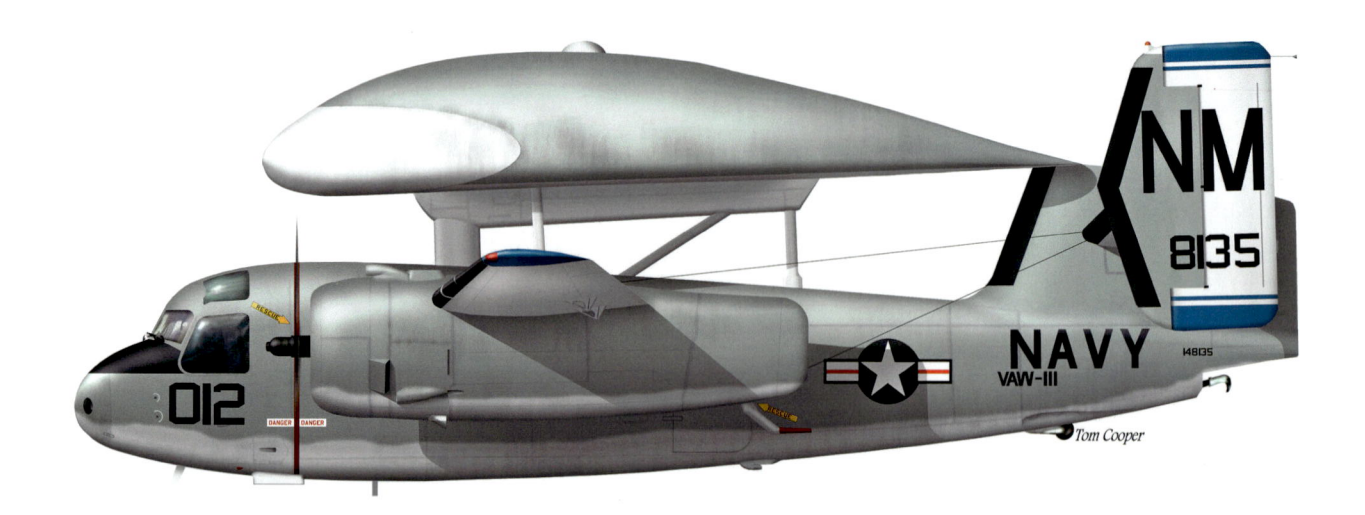

Powerplant
Two Curtiss-Wright R-1820-82A radial engines with maximum output of 1,137kW (1,525hp) at take-off, driving two Hamilton Standard propellers of 3.35m (11ft) diameter

Dimensions
Wingspan, wings spread 22.05m (72ft 4in), wings folded 9.27m (30ft 5in); length 13.82m (45ft 4in); height 5.13m (16 ft 10 in); wing area 47 m^2 (506 sq ft)

Weights
Empty 9,361kg (20,638lb); loaded 11,249kg (24,800lb); maximum take-off 12,066kg (26,600lb)

Performance
Maximum speed 365km/h (227mph) at 1,220m (4,000ft); cruising speed 262km/h (163mph); rate of climb 341m (1,120ft) per minute; service ceiling 4,815m (15,800ft); range 1,625km (1,010 miles)

Avionics
Three AN/ARC-52 UHF radios, AN/ARC-38 VHF radio, AN intercom/CTA-14, AN/ARA-25 automatic UHF direction finder, AN/ARN-12 beacon receiver, AN/APN-22, AN/APX-6B composite IFF transponder device, AN/APX-7 radar signals recogniser, AN/APA-89 code exchanger, AN/ARN-21 TACAN, Collins DF-201 titrator for low-frequency direction, AN/ASA-13 navigation computer

Radar
Hazeltine AN/APS-82 S-band radar, roating at 6.1rpm, and able to detect a target with an RCS of $5m^2$ (54 sq ft) at 507km (315 miles). Two AN/APA-125 monitors, AN/ART-28 data transmitter radar

Production
88

Crew
Pilot, co-pilot and two radar controllers

Operator
United States

Grumman E-2 Hawkeye

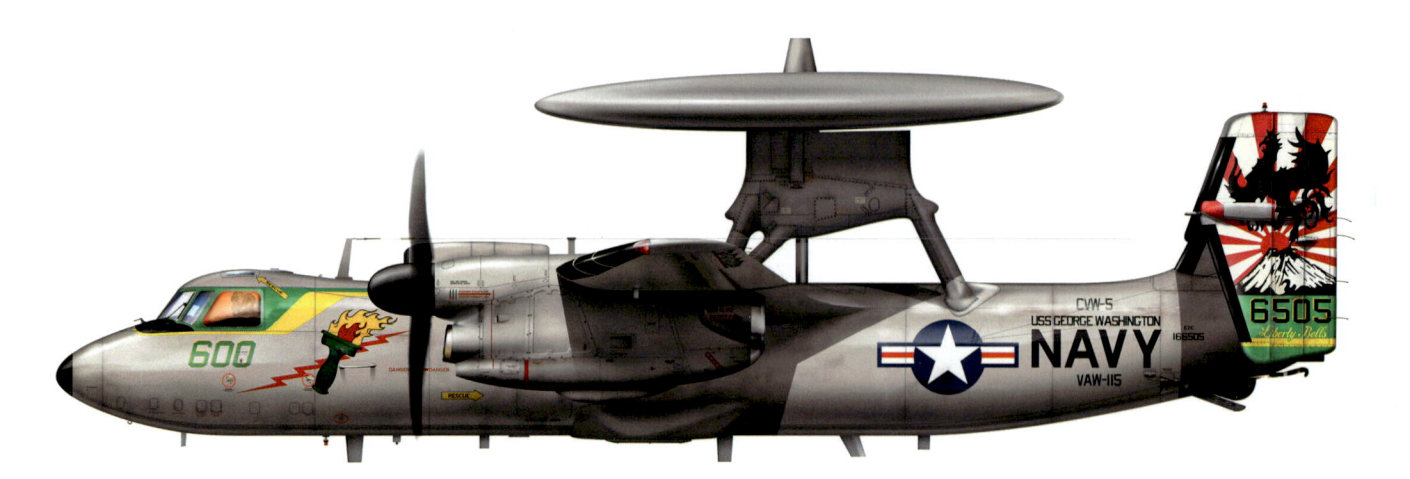

Powerplant
E-2A/B: two Allison T56-A-8/8A turboprops developing 3,020kW (4,050hp) on take-off, driving two four-bladed Aero Products A6441 FN-248 propellers of 4.11m (13ft 6in) diameter
E-2C Basic and Group 0: two Allison T56-A-422/425 turboprops, developing 3,661kW (4,910hp), driving two Hamilton Standard 54460 propellers
E-2C Group I, II, II Plus and Hawkeye 2000: two Allison T56-A-427 turboprops developing 3,803kW (5,100hp); propellers as E-2C
E-2D: two Allison T56-A-427 turboprops developing 3,803kW (5,100hp), driving two eight-bladed Hamilton Sundstrand/Ratier Figeac propellers of 4.67m (15ft 4in) diameter

Dimensions
Wingspan, folded, all models, 8.94m (29ft 4in), unfolded, 24.56m (80ft 7in); length, E-2A/B 17.17m (56ft 4in), E-2C and Hawkeye 2000 17.55m (57ft 7in), E-2D 17.60m (7ft 9in); height, E-2A/B (rotodome raised/retracted) 5.59m (18ft 4in)/5.02m (16ft 6in), E-2C/D (to top of SATCOM antenna) 5.75m (18ft 10.5in); wing area 65.03m² (700 sq ft)

Weight
Empty, E-2A 14,131kg (31,154lb), E-2B 16,336kg (36,014lb), E-2C 17,265kg (38,063lb), late E-2C and E-2D 8,363kg (40,483lb); maximum take-off, E-2A 24,870kg (54,830lb), E-2B 23,133kg (50,999lb), E-2C (early models) 23,556kg (51,933lb), E-2C (late models) 24,687kg (54,425lb), E-2D 25,850kg (56,989lb)

Performance
Maximum speed, E-2A 592km/h (367mph), E-2B 603km/h (374mph), early E-2C 599km/h (372mph), late E-2C/D 626km/h (388mph); cruising speed, E-2A 478km/h (297mph), E-2B and early E-2C 478km/h (297mph), late E-2C 496km/h (308mph), E-2D 480km/h (298mph); service ceiling, E-2A 8,780m (28,800ft), E-2B 8,777m (28,796ft), early E-2C

9,390m (30,800ft), late E-2C/D 11,275m (36,991ft); maximum range, E-2A 3,063km (1,903 miles), E-2B 2,582km (1,604 miles), E-2C and E-2D 4,025km (2,500 miles)

Avionics

E-2A: AN/ASN-36 588-CP mission computer, inertial navigation system (also in E-2B), AN/ARD-13 automatic titrator direction (also in E-2B), AN/APN-122 Doppler navigation system, AN/ARR-66 and AN/ART-66 UHF receiver and transmitter, AN/APN-20 radar altimeter, AN/ASN-50 navigation system (also in E-2B/C), AN/ASH-20 IFF, AN/ASN-27 navigation computer, AN/ASQ-52 airborne tactical data system, (consisting of AN/data ACQ-2 transmitter set, AN/ASW-14A digital data communications system, AN/ARC-80HF radio, AN/AIC-4 interphone also in E-2B, three AN/27 monitors, also in the E-2B)

E-2B: AN/ASH-22 flight data recorder, A24G-13 air data computer, AN/ASW-15 and AN/ASN-31 air navigation computers, AN/APN-22 radar altimeter, AN/ARC-80 and AN/ARC-94 HF radios, two AN/ARC-52 UHF radios (operating with AN/ARR-66 and AN/ART-66 radios to form the 'encrypted voice equipment'), AN/APN-153 Doppler navigation radar (also in the E-2C Group 0 and Groups, I and II), AN/ASQ-58 UHF radios, RT-542/ASQ and ASQ, RT-559/TACAN KY-309/ASQ, RT-988 questioner, AN/ACQ data transmitter

E-2C Basic: AN/AIC-14A intercom (all 'C' models), AN/ARC-51A UHF radio, AN/ARC-104 HF radio, AN/ARN-52 TACAN, AN/ARA-50 automatic titrator direction (all 'C' models), AN/APN-171 radar altimeter (all 'C' models), two AN/AAU-19A altimeters (also in the E-2C Group 0), AN/ASW-15 automatic flight control system (E-2C Group 0 and Group I), UHF AN/ARC-124 digital data-link system (also in E-2C Group 0), AN/ASN-92(V) CAINS inertial navigation system (also in E-2C Groups 0, I and II), AN/ASW-25A automatic landing system for aircraft carriers (also in E-2C Groups 0 and I), AN/ASM-33 in-flight performance monitor (also in E-2C Groups 0 and I), OL-OL CPU KY-533A transponder, 93-77/ASQ radar signal processor (also in Group 0), CP-1085 air data computer, three sets of AN/APA-172 and AN/APQ-179 monitors (of 30.5cm/12in and 12.7cm/5in respectively, also in the E-2C Groups 0, I and II), AN/ALR-59 passive detection system, AN/ASH-22 signal data recorder (also in Group I), AN/ASH-20 flight data recorder

E-2C Group 0: AN/ASW-25 data-link system (compatible with standard Link 4A, also in Group I), AN/APX-72 IFF interrogator (also in Group I), AN/ALR-73 passive detection system (also in Groups I and II), OL-424/ASQ mission computer, RD-576/ASQ digital data recorder (also in Groups I and II), AN/ARN-84 TACAN decoder, AN/ARA-63 automatic titrator, three encrypted AN/ARC-182 VHF radios, two AN/ARC-157 HF radios, three AN/ARC-158 UHF radios (all radios also fitted in Groups I, II, II Plus), AN/ASN-27 navigation computer (also fitted in Groups I, II, II Plus), Mini-DAMA standard satellite communications (also in also fitted in Groups I, II, II Plus)

E-2C Group I: As above, plus AN/ARN-84 TACAN

E-2C Group II: AN/APX-100 IIF transponder (also fitted in Group II Plus), OL-77 mission computer/ASQ, AN/ARC-158 UHF data-link (compatible with Link 4A, 11 and 16 equipment, also in Group II Plus), AN/ARC-34 HF, radio, AN/URC-107(V)5 system JTIDS Class 2 (also in Group II Plus and possibly E-2D), GPS AN/ARN-151(V)2 (also in Group II Plus), Mini LADY navigation, ASM-400 in-flight performance monitor, AN/ASW-25B automatic landing system for aircraft carriers, Loral monitors, MA-1 compass system, AN/APN-154 or AN/APN-202 automatic flight control system, AN/ARN-84 or AN/ARN-118 TACAN, R-1623 radar receiver/decoder receiver for NPA, AN/ARA-63 (also in Group II Plus and E-2D), BAE Systems central air data computer

E-2C Group II Plus/Hawkeye 2000: three AN/UYQ-70 monitors, AN/ALQ-217 countermeasures equipment, CP 2319/ASQ mission computer supplemented by four DEC 2100 Model A500MP, MATT communications terminals, AN/ASN-139 dual navigation suite, CAINS II, UHF AN/ARC-210 radio for satellite communication and data exchange in VHF and (also in E-2D), AN/ASW-50 automatic flight control system, AN/USG-3 CEC air data computer (also in E-2D)

E-2D Advanced Hawkeye: three MOSART monitors, TCAS alert, JTIDS and MIDS communications, GPS Navstar, six AN/ARC-210 VHF and UHF radios (for Link 11 and Link 16), SRT-470 encrypted HF radio, SIRST infrared emission detector, AN/ALQ-117 ESM

Radar
E-2A/B: AN/APS-96 radar with frequency of 406-450MHz,, with horizontal/vertical coverage angles of 360° and 20° respectively with a range of 98 miles (157km) on a contact with a RCS of 1m^2, enclosed in AN/APA-143 rotodome rotating at 6rpm

E-2C Basic: AN/APS-120 radar, operating frequency between 300MHz and 3GHz, in the following models, with identical coverage angles as the AN/APS-96, and detecting similar RCS contacts out to 222km (138 miles). Later replaced by AN/APS-125, with a range of 249km (155 miles) under the same conditions. Both installed within AN/APA-171 rotodome (which also houses the IFF aerial in all E-2C models), turning at 6rpm

E-2C Group 0: AN/APS-138 radar, with maximum range of 402km (250 miles), capable of managing up to 600 contacts

E-2C Group I: AN/APS-139 radar with similar range, but with improved ECCM capabilities and improved surface-search performance. Reported to be able to maintain over 2,400 tracks simultaneously

E-2C Group II: AN/APS-145 digital radar, operating in UHF band (0.3-1GHz, with 16 selectable frequencies), 360° scanning every six seconds, simultaneous tracking of 2,000 targets and trace correlation between 20,000 contacts. Has automatic selection of channel and a frequency best suited to each situation is reported to provide a 40 per cent performance improvement over AN/APS-138 with rotation reduced to 5rpm.

Reported to be able to detect a target with a radar reflection surface of $3m^2$ at around 333km (206 miles) and large contacts at 648km (403 miles), while a small speedboat can be detected at around 138km (85 miles)

E-2D Advanced Hawkeye: AN/APY-9 radar, electronic scanning in elevation and azimuth mechanical/electronic, with BAE Systems ADS-18 integrated interrogator antenna, allowing 360° of coverage and a reported estimated 555km (345 miles) maximum range in three modes of operation: mechanical selectable speed tracking, azimuth tracked electronically and mechanical screening with additional electronic capability, within a zone selectable by the operator

Production
276 (all versions); E-2A (62), E-2B (49, all converted from E-2A), E-2C (200, all versions) and E-2D (75 ordered)

Crew
Five, consisting of pilot, co-pilot and three radar operators

Operators
Egypt, France, Israel, Japan, Mexico, Singapore, Taiwan and United States

Grumman TBM-3W Avenger

Powerplant
One Wright R-2600-20 Cyclone radial engine rated at 1,417kW (1,900hp)

Dimensions
Wingspan 16.52m (54ft 2in); length 12.48m (40ft 11½in); height 5.03m (16ft 6in); wing area 45.52m² (490.02 sq ft)

Weights
Empty 5,394kg (11,893lb); maximum take-off 6,712kg (14,798lb)

Performance
Maximum speed at 5,013m (16,450ft) 418km/h (259mph); cruising speed 231km/h (144mph); rate of climb 356m (1,085ft) per minute; service ceiling 8,868m (28,500ft); range 1,359km (844 miles)

Avionics
AN/ART-13 IFF, AN/ARC-1 VHF transceiver, AN/APX-1AM transponder, RL-7 intercom, AN/ARC-18 radio repeater, AN/ART-2 relay antenna, AN/ART-22 relay transmitter (AN/ART-26 in some foreign aircraft), AN/ART-28 relay receiver, AN/ARW-35 receiver, AN/APN-1 radio altimeter, AN/APA-56 30.5cm (12in) displays (used with AN/APA-57 ground position indicator and AN/APA-81 group position indicator)

Radar
AN/APS-20 radar (see Douglas AD-3W, 4W and 5W Skyraider)

Production
TBM-3W (40), TBM-3W2 (156, optimised for ASW operations)

Crew
One pilot and one radar technician

Operators
France, Japan, the Netherlands and United States

Gulfstream G550 CAEW

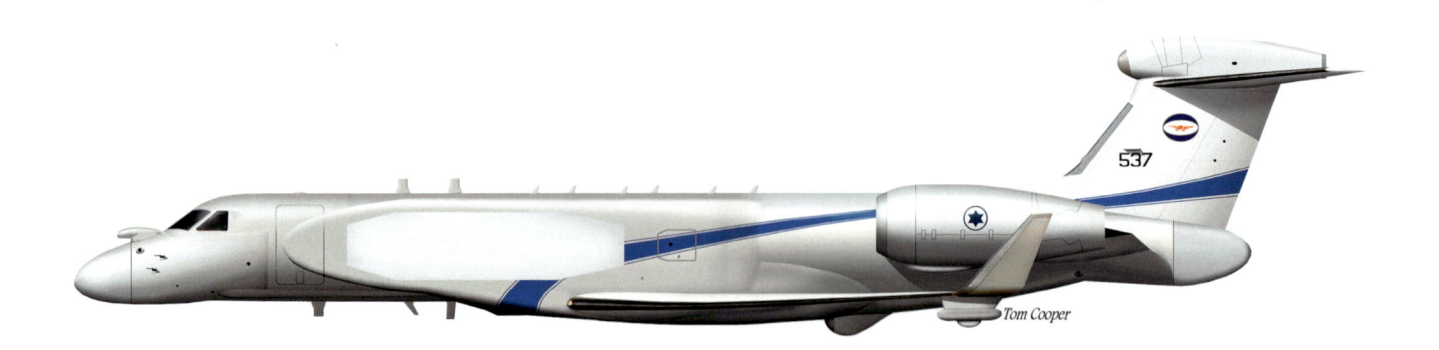

Powerplant
Two Rolls-Royce BR710-48 turbofans, rated at 6,980kg (15,550lb) thrust

Dimensions
Wingspan 28.50m (93ft 6in); length 29.40m (96ft 5in); height 7.90m (25ft 10in); wing area 105.60m² (1,136 sq ft)

Weights
Not known

Performance
Maximum speed around 938km/h (583mph); cruising speed around 753km/h (468mph); service ceiling 12,496m (41,000ft); maximum ceiling 15,544m (51,000ft); range 12,509km (7,772 miles); endurance around 9 hours

Avionics
SPZ-8500 flight director system, Primus II Epic avionics system, Elta EL/L-8382 ESM/ELINT suite, Elta EL/K-1891 satellite and data-link communications system, Elta EL/M-2610 self-protection system with RWR, MAWS, DIRCM, chaff and flare dispensers

Radar
Elta EL/M-2085 Doppler/AESA, L-band/S-band radar (1-2GHz and 2-4GHz); system mounted in nose and tail radomes and fuselage cheek conformal panels; integrated IFF, capable of determining target trajectory within four seconds of detection. Up to 100 targets can be tracked simultaneously and over a dozen air-to-air interceptions or air-to-ground attacks can be managed at a time. Modes: automatic tracking, scanning area selection, high agility targets detection. Radar range: 370km (230 miles)

Appendix I

Production
8

Crew
Eight+, consisting of two flight crew and around six or more mission crew

Operators
Italy (on order), Israel and Singapore

IAI Condór

Powerplant
Four Pratt & Whitney JT3D-7 turbofans each rated at 9,491kg (21,000lb) thrust

Dimensions
Wingspan (44.42m (145ft 9in); length 46.61m (152ft 11in); height 12.93m (42ft 5in); wing area 283.4m² (3,050.5 sq ft)

Weights
Empty 77,996kg (171,948lb); maximum take-off 151,953kg (334,953lb)

Performance
Maximum speed 851km/h (529mph); cruising speed 678km/h (421mph); service ceiling 11,582m (38,000ft); maximum range 8,140km (5,058 miles)

Avionics
Navigation/communication suite as commonly installed on Boeing 707-385C, augmented by EL/M-2610 IFF, EL/S-8610 central computer, EL/L-8300 airborne SIGINT system that incorporates EL/K-7032 COMINT system, EL/L-8312A ELINT system and EL/L-8350 command and analysis station. CSM/COMINT receives in UHF, VHF and HF, rapidly searching for airborne, shipborne or ground communications signals of interest. Selected radio nets can be monitored for signal activity. A DF capability locates targets. Communicates via data-link with air defence HQ

Radar
Elta EL/2075, L-band phased-array radar conformally mounted on both sides on forward fuselage cheeks and large bulbous nose radome providing approximately 180° coverage; electronically steered beams consisting of 768 elements that can be controlled individually; track initiation is achieved in two to four seconds as compared to 20 to 40 seconds with a rotodome-mounted radar. Solid-state phased-array IFF system able to perform interrogation, decoding, target detection and tracking IFF data is auto-

Appendix I

matically correlated with the phased array radar. ESM/ELINT system receives, analyzss and locates radar signals, and combines high sensitivity with high probability of intercept; system uses narrow-band super-heterodyne receivers and wide-band instantaneous frequency measurement (IFM) techniques; bearing accuracy for all received signals is achieved through differential time of arrival (DTOA) measurements, system also collects and analyses ELINT data. The Phalcon system allows up to 12 simultaneous vectored intercepts to take place simultaneously

Production
2

Crew
Not known

Operators
Chile and Israel (IAI)

Kamov Ka-31

Powerplant
Two Izotov (NPP Klimov) TV3-117VMA turboshafts each rated at 1,641kW (2,200shp)

Dimensions
Length 12.5m (41ft), rotor diameter 15.9m (52ft 1in), fuselage width 3.8m (12ft 5in), height 5.6m (18ft 4in)

Weights
Empty 5,520kg (12,170lb); maximum take-off 12,500kg (27,560lb)

Performance
Maximum speed at sea level 280km/h (174mph); loiter speed 100-120km/h (62-75mph); service ceiling around 3,048m (10,000ft); maximum range 600km (372 miles), loiter duration 2 hours 30 minutes

Avionics
Russia: PKP-77 flight director, PNP-72-15 gyro horizon, IK-VSP-V1-2 altitude and speed data system, DISS-32 Doppler speed/drift sensor, A-036A radio altimeter, ARK-22 automatic direction-finder, A-723 LORAN, A-380 SHORAN, AGR-74V-10 artificial horizon, range and speed indicators. Ka-31R known to be equipped with BKS-252 radio communication suite

China: similar to the basic model, except for the following: NK-37DME navigation/flight system, A-052 radio altimeter, ARK-25 ADF, AGR-81 artificial horizon, INP-R flight course indicator, AT-E data-link

India: similar to basic and Chinese models, apart from the Kronstadt Abris GPS and Bharat Electronics Link IIA data-linkwith a data rate of 2,400kbps

Radar
Russia: E-801 decimeter-waveband radar with an under-fuselage flat antenna that deploys through 90° from flat stowed position, rotating through 360° after take-off, rotating at 6rpm; landing gear retracts upwards to prevent interference. Once radar has been selected and turned on, the system employs fully automatic and semi-automatic target detection, target location, tracking and trajectory data on air targets flying below helicopter's altitude and surface targets, all of which is transmitted automatically to shipboard or land based command centres.
Maximum surveillance radius 100–200km (62–124 miles) for fighter-sized targets, 124 miles (200km) for surface vessels (from an altitude of 3,000m/9,840ft), up to 40 airborne or surface targets can be tracked simultaneously

China and India: slightly downgraded E-801E radar export version

Production
20+

Crew
One pilot and one radar technician

Operators
China, India and Russia

Lockheed Neptune AEW.Mk 1

Powerplant
Two Wright R-3350-30W 18-cylinder two-row radial engines, each rated at 2,425kW (3,250shp)

Dimensions
Wingspan 31.7m (104ft); length 23.85m (78ft 3in); height 8.81m (28ft 11in); wing area 92.9m² (1,000 sq ft)

Weights
Empty 18,098kg (39,900lb); maximum take-off 35,312kg (77,850lb)

Performance
Maximum speed 489km/h (303mph); rate of climb 366m (1,200ft) per minute; service ceiling 7,620-9,144m (25,000-30,000ft); range 6,440km (4,607 miles); maximum duration 15.6 hours

Avionics
AN/AIO-5B intercom, A/ART-13 radios, AN/APX-6 or -7 IFF, AN/ARC-1 VHF radio, A/ART-22 (later AN/ART-26), SSR Mk 4, A/ARA-6 direction-finder, AN/APN-22 radio altimeter, AN/APN-70 LORAN, AN/AIC-5 receiver, AN/ARP-15A liaison receiver, AN/ARC-27 UHF radio, APN-12 radio beacon, AN/APA-9B or 11 ECM, AN/ARN-21 DME, AN/APA-69 ECM direction-finder, AN/ARN-14A ECM omnidirectional finder, AN/APA-57B ground position indicator, AN/APA-81 12.7cm (5in) PPI displays

Radar
AN/APS-20B radar, with 2MW power source

Production
5 aircraft converted from P2V-5s

Crew
Six to eight, depending on sortie

Operator
United Kingdom

Lockheed Martin P-3 AEW

Powerplant
Four Allison T-56-A-14 turboprops developing 3,661kW (4,910shp) and driving four Hamilton Standard propellers of 4.11m (13ft 6in) diameter

Dimensions
Wingspan 30.37m (99ft 8in); length 35.61m (116ft 10in); height 10.27m (33ft 8½in); wing area 120.77m² (1,300 sq ft).

Weights
Empty 36,565kg (80,612lb); maximum take-off 61,234kg (135,000lb)

Performance
Maximum speed 481km/h (299mph) at 4,572m (15,000ft); patrol speed 333km/h (207mph); service ceiling 7,925m (26,000ft); range 7,165km (4,452 miles)

Avionics
AN/TPX-54 IFF (with improved capabilities in updated aircraft), five AN/ARC-210 Flexcomm II V/UHF radios, two RT-5000 and OTAR FM with receivers/transmitters, two UHF satellite LST-5 d communication radios, three KYV-5 self-encrypting voice and radio, three KY-58 voice self-encrypting, AN/ARC-174 HF radio with beyond-the-horizon range, COTHEN network (which connects 235 ships, aircraft and all US government agencies engaged in anti-narcotics combat), four Q-70 workstations with three colour 51cm (20.1in) flat-screen monitors (plus a 33cm/13in tactical monitor in the cockpit that repeats the same image seen on the monitor of the radar operator, controlled by the pilot with a joystick), five CDU-800 communication controls, double Litton 92 inertial navigation platforms, TWR-850 weather radar, dual-channel autopilot, EFIS display system for engine instruments, dual VIR-432 VOR, dual DME-442 nav kit, dual ADS-3000 air data system, ALT-4000 radar altimeter, GH-3000/ADC-3000 flight management system with associated CMA-900 GPS, T2CAS IFF Mode S anti-collision equipment

Radar
AN/APS-145 digital radar, operating in UHF band (0.3-1GHz, with 16 selectable frequencies), with 360° scanning every six seconds, simultaneous tracking of 2,000 targets and trace correlation between 20,000 contacts. Has automatic selection of channel and frequency best suited to each situation, rotates at 5rpm.
Reported to be able to detect a target with a radar reflection surface of $3m^2$ at around 333km (206 miles) and large contacts at 650km (403 miles), while a small speedboat can be detected at around 138km (85 miles)

Production
8

Crew
Eight or more, consisting of three flight crew (pilot, co-pilot, flight engineer) and five or more radar operators

Operator
United States

Lockheed WV and EC-121 Warning Star

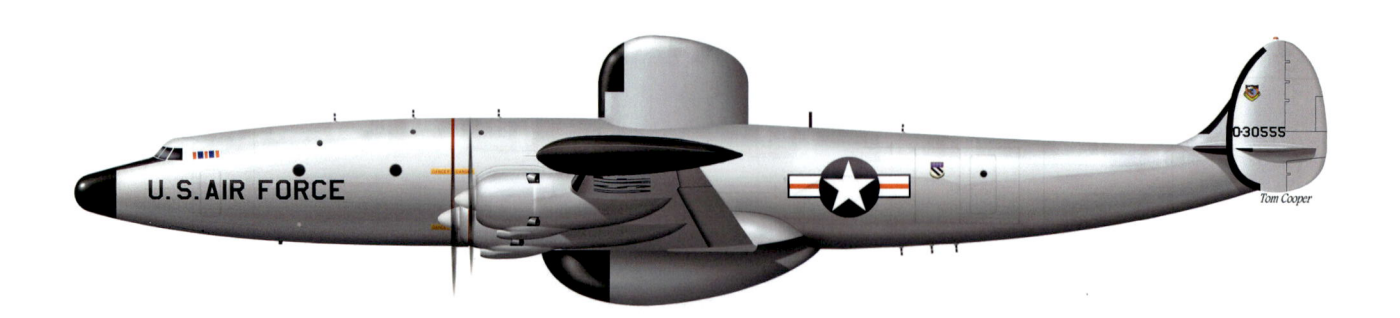

Powerplant
Four Pratt & Whitney R-3350-34 Double Wasp (or R-3350-42 Turbo-Compound) piston engines each developing 2,536kW (3,400shp), driving Hamilton Standard 6903A-0 propellers of 4.62m (5ft 2in) diameter

Dimensions
Wingspan 37.62m (123ft 5in); length 35.41m (116ft 2in); height 8.23m (27ft); wing area 153.66m^2 (1,654 sq ft)

Weight
EC-121: empty 36,565kg (80,611lb); maximum take-off 65,136kg (143,600lb)

Performance
EC-121: maximum speed 516km/h (321mph); cruising speed 386km/h (240mph); rate of climb 258m (845ft) per minute; service ceiling 6,280m (20,600ft); normal range 7,400km (5,297 miles)

Avionics
WV-2: AN/ARC-27 UHF transmitter/receiver, AN/ARC-38 HF transmitter/receiver, AN/ARC-1 VHF transmitter/receiver, AN/ARR-41 HF receiver, R-23A/ARC-5 LF receiver, AN/APX-6 (or-6B), AN/ARN-21 transponder navigation set, AN/ARN-14 VOR, AN/ARN-18 glide slope indicator, AN/ARN-6 compass radio, AN/APN-70 LORAN receiver, AN/APN-22 altimeter radar (or AN/SCR-71 for meteorological missions), AN/ARN-12 marker beacon receiver, AN/APR-13 ECM receiver, AN/APA-74 ECM signal analyser, AN/APA-81 radar indicator group, AN/ALA-2 scenic ECM indicator, AN/APR-9B ECM receiver, AN/ARA-25 UHF direction designator, AN/APX-7 IFF; AN/ARR-27A transmission receiver radar, AN/ART-28 transmitter/receiver radar, AN/APA-57 ground position indicators

EC-121D: two AN/ARC-85 UHF radios, AN/APX-25 IFF (later AN/APX-49 and AN/APX-101), AN/ARN-14 VOR/TACAN, AN/ARN-18 glide slope indicator, AN/APS-42 weather radar, seven AN/ARC-27 UHF radios, AN/ARN-21 TACAN, AN/APN-70 long-range browser, AN/ARN-6 ADF, AN/ARN-25 UHF direction designator, AN/SCR-718 and AN/APN-27 radar altimeters, ALRI data transmission device and adjustable antenna (model H), five AN/APA-56 display screens, five AN/APA- 57 ground position indicators

Radar
WV-2: AN/APS-20B, S-band (2-4GHz), 360° rotation, 6 or 10rpm selectable tracking velocity, with designed maximum range of 370km (230 miles). Maximum range was rarely achieved and then only with very large targets.
EC-121D/Q: AN/APS-95 radar, P-band (406-450MHz), 360° rotation, NATO B-band with a peak power of 3MW; Texas Instruments AN/APS-45 height-finding radar, X-band (7.0-11.2GHz), capable of detecting contacts flying up to 12,192m (40,000ft) with a maximum range of 222km (159 miles), the later AN/APS-103, had a maximum range of 296km (212 miles) on the EC-121Q

Production
WV-1 (2), WV-2 (142), WV-3 (8)
RC-121/EC-121 (83)

Crew
Between 17 and 27 depending on variant and mission requirements

Operator
United States

Saab 340 AEW&C and Saab 2000 AEW&C

Powerplant
S100B Argus: two General Electric CT7-9B turboprops each rated at 1,394kW (1,870hp), driving two Dowty four-blade constant speed propellers of 3.35m (11ft) diameter
Saab 2000 AEW&C: two Rolls-Royce AE 2100A turboprops each rated at 3,096kW (4,152hp), driving two six-bladed propellers of a 3.81m (12ft 6in) diameter

Dimensions
S100B Argus: wingspan 21.44 m (70ft 4in); length 19.73m (64ft 8¾in); height 6.97m (22ft 10½in); wing area 41.81m² (450 sq ft)
Saab 2000 AEW&C: wingspan 24.76m (81ft 3in); length 27.28m (89ft 6in); wing area 55.7m² (599.55 sq ft)

Weights
S100B Argus: empty 8,225kg (18,133lb); maximum take-off 13,155kg (29,000lb); Saab **2000 AEW&C:** empty 14,500kg (31,966lb); maximum take-off 23,000kg (50,706lb)

Performance
S100B Argus: maximum speed 463km/h (288mph); patrol speed 300km/h (186mph); optimum cruising speed 448km/h (278mph) at 7,260m (25,000ft); service ceiling 7,620m (25,000ft); range 2,858km (1,776 miles)
Saab 2000 AEW&C: maximum speed 657km/h (408mph); service ceiling 9,144m (30,000ft); range 4,287km (2,664 miles)

Avionics
S100B Argus: Rockwell Collins APS-85 automatic flight control system, Sundstrand Lockheed flight data recorder and ground proximity warning, Honeywell H764G inertial/GPS navigator, TILDES standard instrument landing, FR-JAS suite of HF/VHF/UHF radios, Mk 12 IFF/SSR Mode 1 to 4 interrogator, Saab.Ericsson consoles with 19.5in LCD monitor and two 10.4in panels

Saab 2000 AEW&C: WXR 840 weather radar, Collins Pro Line 4 integrated navigation system, TRIALS 300 electronic warfare suite (including HES-21, with electronic support measures with 360° coverage, detecting emissions between 2.7-40GHz), self-protection system with thermal decoys derived from the MAW-300 (release alert and missile approach, capable of detecting eight projectiles fired at the aircraft), laser warning based on LWS-310

Radar
S100B Argus: dorsal-mounted Ericsson PS-890 Erieye radar pulse-Doppler SLAR with fixed active-array antenna in S-band (3.1-3.3GHz) AESA radar, with 300° area coverage. Antenna pod 9m (29ft 6in) long and weighs more than 900kg (1,984lb). Radar range coverage typically 350km (217 miles) against fighter-sized targets, 150km (93 miles) against low-flying cruise missiles and 300km (186 miles) against surface ships. Able to track 300 simultaneous contacts and vector at least six aircraft simultaneously. Operates in adaptive modes, target tracking, surveillance-while-search, extended air alert, primary and secondary air surveillance and maritime surveillance

Saab 2000 AEW&C: PS-890 Erieye radar coverage increased to 360°

Production
S100 Argus (6), Saab 2000 AEW&C (4)

Crew
S100B Argus: two flight crew, up to three mission crew
Saab 20000 AEW&C: two flight crew, up to five mission crew

Operators
Greece (Saab 340 AEW&C), Pakistan (Saab 2000 AEW&C), Saudi Arabia (Saab 2000 AEW&C), Sweden (S100 Argus), Thailand (Saab 340 AEW&C), United Arab Emirates (Saab 340 AEW&C)

Shaanxi KJ-200, ZDK-03 and Y-8J

Powerplant
KJ-200: four uprated Zhuzhou WoJiang-6C (WJ-6C) turboprops rated at 3,805kw (5,100shp) and driving JL-4 six-bladed composite propellers
Y-8J: four SAEC (Zhuzhou Wojiang) WJ6A turboprops each rated 3,126kW (4,192hp) and driving Baoding four-bladed propellers

Dimensions
KJ-200: wingspan 40m (131ft 2in); length 36m (118ft 1in); height 11.3m (37ft 1in); wing area 121.86m² (1,311.7 sq ft)
Y-8J: wingspan 38m (124 ft 6in); length 4.02m (111ft 7in); height 11.6m (38ft); wing area 121.86m² (1,311.7 sq ft)

Weights
KJ-200: empty in excess of 39,000kg (85,980lb); maximum take-off in excess of 65,000kg (143,300lb)
Y-8J: empty 35,488kg (78,236lb); maximum take-off 61,000kg (134,480lb)

Performance
KJ-200: maximum speed 650km/h (404mph); cruising speed 550km/h (342 miles); rate of climb not known; operational ceiling 10,100m (33,136ft); range 5,000km (3,107 miles)
Y-8J: maximum speed 662km/h (411mph), cruising speed 550km/h (342mph); rate of climb 600m (1,970ft) per minute; operational ceiling 10,400m (31,699ft); range 5,620km (3,492 miles)

Avionics
KJ-200: additionally equipped with full C3I centre, and a new integrated digital avionics system based on ARING429 and RS422 databuses. ESM antennas in fairings at the wingtips and on top of the tailfin housing

Y-8 (generic): Rockwell Collins VHF-42B and HF-9000 radios, TDR-94 ATC transponder, UNS-1K flight management system, Rockwell Collins VOR-432, DME-442, AH-85E, AHRS and EFIS-86E, Honeywell ED-55 flight data recorder

Radar
KJ-200: 'balance beam' AESA electronically steered phased array radar developed by the 38th Institute, similar in configuration and appearance to the Ericsson Erieye, with a purported range of 300-450km (186-280 miles). Radar is unable to scan the forward and rear sectors of the aircraft and thus lacks full 360° coverage, apparently compensated by two additional radomes located at the nose and tailcone which may house additional antennas

Y-8J: Racal Skymaster 360° azimuth-sweep radar mounted in radome under the nose with range estimated to be 370km (230 miles)

Production
12 KJ-200, 4 ZDK-03, 4 Y-8J

Crew
Eight or more, consisting of pilot, co-pilot, navigator, engineer and radio operator, plus three or more mission crew depending on variant

Operators
China (KJ-200, Y-8J), Pakistan (ZDK-03)

Tupolev Tu-126

Powerplant
Four Kuznetsov NK-12MV turboprop engines, each developing 11,185kW (15,000hp) on take-off, and driving two AV-60K contra-rotating propellers of 5.6m (18ft 4in) diameter

Dimensions
Wingspan 51.4m (168ft 7in); length (with aerial refuelling probe and ECM fairing) 57.9m (189ft 11in); height 15.5m (50ft 10in); wing area 311.1m² (3,349sq ft)

Weights
Empty 103,000kg (227,076lb); normal take-off 155,172kg (342,090lb) maximum take-off weight 171,000kg (376,990lb)

Performance
Maximum speed 790km/h (490mph); service ceiling 10,700m (35,100ft); range 7,000km (4,347 miles); endurance 10.2 hours

Avionics
AP-15 autopilot, Put-1 semi-automatic navigation system, long-range navigation system, RSBN-25 radio-navigation system, R-Rubin-1Sh radar bombing/navigation aid, ARK-11 (main) and ARK-U2 (reserve) automatic direction-finders, GPK-52 gyrocompass, RV-4 (RV-a) radio-altimeter, AGD-1S artificial horizons, VD-20 combined airspeed indicator/barometric altimeter, UUT arfagem angle indicator, EUP-53 vertical tilt indicator, Model 13-20ChP timers, VAR-30m vertical speed indicator, SSN-1 dynamic pressure indicators, MS-1 Mach speed meter, DAK-FB-5 remote celestial compass, AK-53 celestial compass, SP-1m Sextant periscope, SPU-7 intercom, SGU-16 (external) radios, US-8 air-to-air and air-to-ground short-range radio, 1-RSB-70 or R-807 high-frequency radios air-to-air and air-to-ground long-range radio (or Mikron in the later versions of the Tu-126; such radios, along with other components formed the Kristal-L data-link, able to manage 10 air intercepts), SRS-6A-7 ELINT devices and SRS and SRS signals intelligence, 2B-126 ASO chaff dispensers (in the aircraft 61M601, 65M611, 66M613 and 66M621)

Radar
Liana radar (NATO codename Flat Jack), capable of 360° coverage, housed in an RA-10 rotodome, turning at 10rpm. Capable of detecting 14 targets at an altitude higher than the Tu-126. Could detect MiG-17-size aircraft at 161km (100 miles), a tactical bomber-size target such as an Il-28 at 322km (200 miles), a strategic bomber target such as the Myasishchev M-4 /3M bomber at 483km (300 miles) or a cruiser-sized surface contact at 644km (400 miles), with the Tu-126 flying at an altitude of 2,000-5,000m (6,562-16,404ft)

Production
9

Crew
12

Operator
Soviet Union

Vickers Wellington Mk IC and Mk XIV

Powerplant
Mk IC: two Bristol Pegasus XVIII radial engines, each rated at 783kW (1,050hp)
Mk XIV: two Bristol Hercules XVII radial engines, each rated at 1,294kW (1,735hp)

Dimensions
Mk IC: wingspan 26.26m (86ft 2in); length 19.68m (64ft 7in); height 5.31m (17ft 5in); wing area 78.1m² (840 sq ft)
Mk XIV: as above, except height 5.35m (17ft 6in)

Weights
Mk IC: empty 8,435kg (18,556lb); maximum take-off 2,955kg (28,500lb)
Mk XIV: empty 8,459kg (18,648lb); maximum take-off 13,381kg (29,500lb)

Performance
Mk IC: maximum speed at 4,570m (14,993ft) 380km/h (236mph); cruising speed 315km/h (195mph); service ceiling 5,454m (17,893ft); maximum range 4,106km (2,551 miles)
Mk XIV: maximum speed at 4,570m (14,993ft) 400km/h (247mph); cruising speed unknown; rate of climb unknown; service ceiling 5,000m (16,404ft); maximum range 3,259km (2,025 miles)

Avionics
Type B or A.1134 intercom, Lorenz ILS (with R.1124A/R.1125A receivers), Marconi GP transceiver radio, Gee Mk 2 navigational device, IFF device with R.3061/3090 receivers; ARI.5002, ARI.5022 and ARI 5033 radios

Radar

Modified ASV Mk II Type R3039A radar reciever, peak pulse power rating 100kW. A 23cm (9in) PPI cathode ray tube, a radial time base was rotated in synch with the antenna rotation. Amplitute modulation was used to brighten the time base and a received contact against the range markers. Calibration markers were displayed at 8km (5 miles) intervals by a ringing circuit. The displayed ranges could be varied between 8-80km (5-50 miles). The PPI screen was enclosed by a calibrated scale around its circumference with zero degrees at the 12 o'clock position. Antenna 10 element horizontally polarised High Gain Yagi narrow beam functioning at 176MHz. Comprising of a driven folded dipole element, a reflector and eight director parasitic elements projecting from a metal covered streamline casing with the dimensions of 38cm (15in) wide at the centre tapering to 23cm (9in) at the ends, and 4.6m (15ft 1in) long. The antenna was attached at the centre to a vertical shaft that projected through the roof of the aircraft. The shaft was rotated inside the fuselage via a step-down gearbox and belt driven by a 24V DC motor. The antenna rotated at 25rpm. The beam width was around 25° of arc. Power to the radar equipment was provided from a Type U 1.2 Kva, 80V, 1,000Hz generator, which was coupled directly to the port engine. Voltage control was provided by a Type 5 voltage control panel. The overall weight of the radar installation was 318kg (701lb).

Production

One Mk IC converted to ACI; apparently only one Wellington Mk XIV was adapted to carry out Operation Vapour sorties

Crew

Five flight crew plus two radar technicians

Operator

United Kingdom

Westland Sea King AEW.Mk 2/ASaC.Mk 7 and Sikorsky SH-3(AEW) Sea King

Powerplant
Mk 2/2A and Mk 7: two Rolls-Royce Gnome H1400-1 turboshafts, with 1,238kW (1,660shp) maximum unitary power for two and a half minutes

Dimensions
Fuselage length 17.01m (55ft 9¾in), with rotors turning 22.15m (72ft 8in), with main and tail rotors folded 14.40m (47ft 3in); width (with flotation bags fitted) 4.98m (16ft 4in); main rotor diameter 18.90m (60ft 4in); height (with rotors turning) 5.13m (16ft 10in), (with rotors stationary) 4.085m (15ft 11in)

Weights
Empty 7,417kg (16,352lb); maximum take-off 9,752kg (21,500lb)

Performance
Maximum speed, Mk 2/2A 186km/h (115mph), Mk 7 226km/h (140mph); cruising speed at sea level 204km/h (127mph); service ceiling around 3,048m (10,000ft); range around 1,230km (764 miles)

Avionics
Mk 2 and Mk 2A: Racal Doppler 91, Racal RNS-252 tactical navigation computer, Smiths/Honeywell AN/APN-198 radar altimeter, Chelton 700 guidance system, Rockwell Collins AN/ARC-182 and HF HF-9000 VHF/UHF radios, Cossor 3570/Jubilee Guardsman IFF, Ferranti FIN 1110 inertial navigation system, Ecko AW 391 navigation radar, W15 (ARI 23283) VHF/FM radio, ARI 23099/25 communications control system, ARI 5954 transponder, ARI 23360 UHF guidance system, GM7B compass system, ARI 23159 'telebrief' system, AD3400 (ARI and ARI 1/23338 23338/2) encrypted in VHF/UHF radios, 618T-3 (ARI 23090/9) HF radio, Mk 31 autopilot, ARI 23331 broadband/7 secure

Appendix I

voice system, D403M (ARI 23159) radio system of UHF, AN/ARC-164(V) (ARI/23315 system JTIDS, 5) UHF, Autocat UHF radio communication system, RT 221 (ARI 23239, all equipment in Mk 7), VHF radio/INGPS system, PTR 446 (ARI & 5970 5970/5) IFF, Violet Picture, inertial VHF guidance system (ARI 18120, these three equipment only in Mk 5/6), Racal MIR 2 Orange Crop (also in Mk 7)

ASaC.Mk 7: Two AN/ARC-164 encrypted UHF radios, Litton LN-100g contact ally/ MkXII hostile handle inertial/GPS navigator, Rockwell radio IDS-2000 (data-link system Link 16-JTIDS, able to manage 300 contacts)

Radar

Mk 2 and 2A: ARI 5930/3 Searchwater radar I, L/low-bandwidth J band (8-12.5GHz) able to track 40 surface targets simultaneously (32 deposited into memory) and 16 aerial targets (100 stored in memory), in addition to controlling six interceptions a a time, in full 360° cover. Has the following operating modes: air-to-air surveillance (with pulse Doppler), anti-ship, anti-submarine, side scan, synthetic aperture, classification of targets, mark friend-enemy, search and rescue, meteorological research, detection above and below the flight level. At an altitude of 3,048m (10,000ft) is able to detect a snorkel of a submarine at 82km (51 miles), a speedboat at 111km (68 miles), a fighter at 129km (80 miles), a bomber at 298km (185 miles), and a medium-sized warship at 240km (149 miles)

ASaC.Mk 7: Cerberus system, which can detect 250 contacts in conjunction with the Searchwater radar 2000 AEW, operating on the same frequency, with pulse-Doppler receivers/transmitters and MTI. Modes of operation include: air-to-air, surface surveillance, target classification, weather detection, with 'normal' synthetic aperture inverted and highlighting an area, anti-ship and anti-submarine combat mode. It is able to perform simultaneous detection of more than 100 targets and control up to nine interceptors via data link. Maximum radar range around 370km (230 miles)

Production
18 (Sea King AEW.Mk 2/ASaC.Mk 7), 3 [SH-3(AEW)]

Crew
One pilot and two observer/radar operators

Operators
Spain [SH-3H(AEW)], United Kingdom (Sea King AEW.Mk 2/2A, ASaC.Mk 7)

APPENDIX II

Service Time Lines and Radar Types and Ranges

Graph 1 (pp242–243) ▶
The following graphic represents every AEW and AEW&C aircraft to have entered full-scale service together with the spans of their respective operational lives.

Graph 2 (p244) ▶ ▶
This graphic depicts the respective range and altitude detection limits achieved by the primary radar sensors on board the various AEW and AEW&C aircraft to have entered full-scale service up until today.

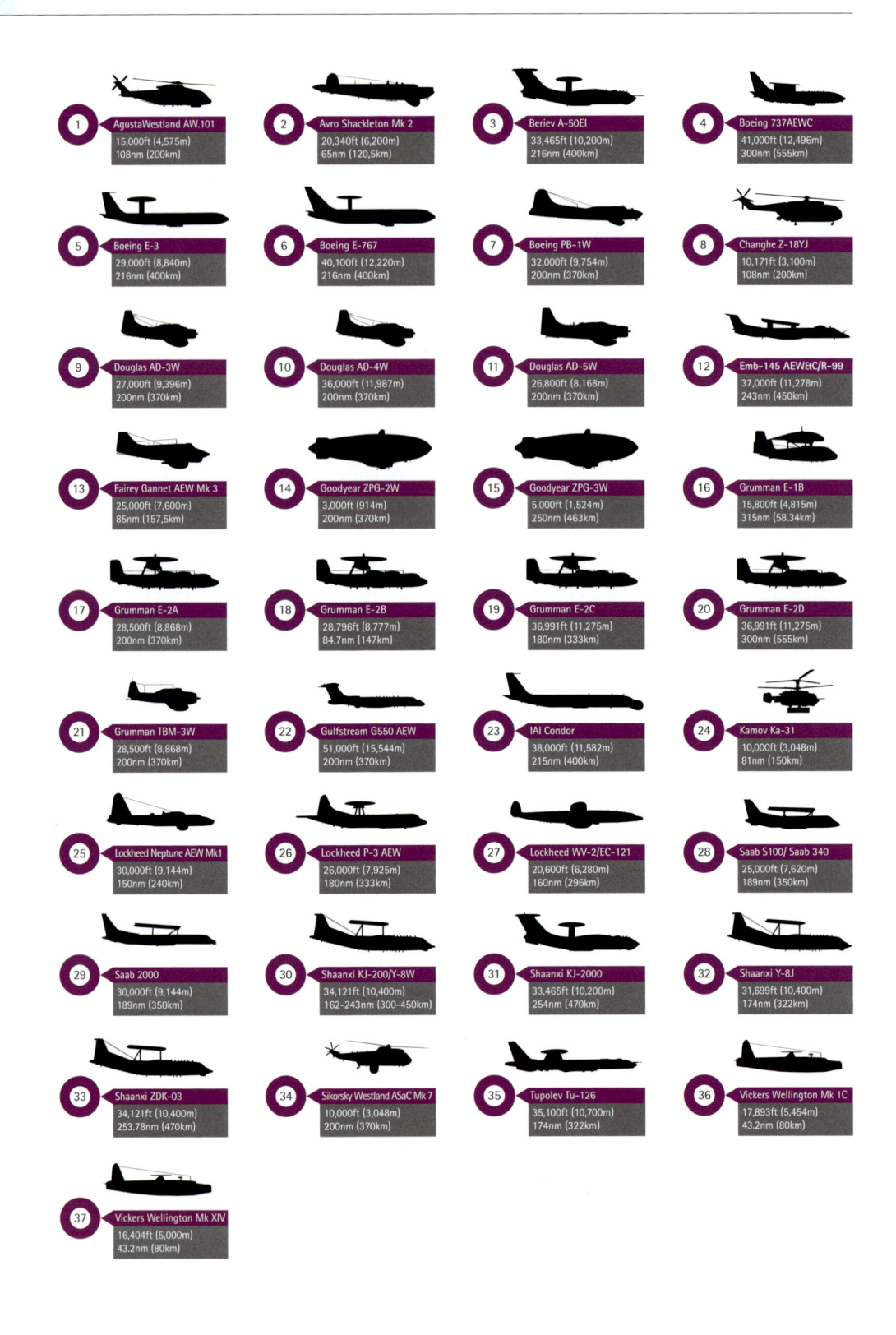

#	Aircraft	Altitude	Range
1	AgustaWestland AW.101	15,000ft (4,575m)	108nm (200km)
2	Avro Shackleton Mk 2	20,340ft (6,200m)	65nm (120,5km)
3	Beriev A-50EI	33,465ft (10,200m)	216nm (400km)
4	Boeing 737AEWC	41,000ft (12,496m)	300nm (555km)
5	Boeing E-3	29,000ft (8,840m)	216nm (400km)
6	Boeing E-767	40,100ft (12,220m)	216nm (400km)
7	Boeing PB-1W	32,000ft (9,754m)	200nm (370km)
8	Changhe Z-18YJ	10,171ft (3,100m)	108nm (200km)
9	Douglas AD-3W	27,000ft (9,396m)	200nm (370km)
10	Douglas AD-4W	36,000ft (11,987m)	200nm (370km)
11	Douglas AD-5W	26,800ft (8,168m)	200nm (370km)
12	Emb-145 AEW&C/R-99	37,000ft (11,278m)	243nm (450km)
13	Fairey Gannet AEW Mk 3	25,000ft (7,600m)	85nm (157,5km)
14	Goodyear ZPG-2W	3,000ft (914m)	200nm (370km)
15	Goodyear ZPG-3W	5,000ft (1,524m)	250nm (463km)
16	Grumman E-1B	15,800ft (4,815m)	315nm (58.34km)
17	Grumman E-2A	28,500ft (8,868m)	200nm (370km)
18	Grumman E-2B	28,796ft (8,777m)	84.7nm (147km)
19	Grumman E-2C	36,991ft (11,275m)	180nm (333km)
20	Grumman E-2D	36,991ft (11,275m)	300nm (555km)
21	Grumman TBM-3W	28,500ft (8,868m)	200nm (370km)
22	Gulfstream G550 AEW	51,000ft (15,544m)	200nm (370km)
23	IAI Condor	38,000ft (11,582m)	215nm (400km)
24	Kamov Ka-31	10,000ft (3,048m)	81nm (150km)
25	Lockheed Neptune AEW Mk1	30,000ft (9,144m)	150nm (240km)
26	Lockheed P-3 AEW	26,000ft (7,925m)	180nm (333km)
27	Lockheed WV-2/EC-121	20,600ft (6,280m)	160nm (296km)
28	Saab S100/ Saab 340	25,000ft (7,620m)	189nm (350km)
29	Saab 2000	30,000ft (9,144m)	189nm (350km)
30	Shaanxi KJ-200/Y-8W	34,121ft (10,400m)	162-243nm (300-450km)
31	Shaanxi KJ-2000	33,465ft (10,200m)	254nm (470km)
32	Shaanxi Y-8J	31,699ft (10,400m)	174nm (322km)
33	Shaanxi ZDK-03	34,121ft (10,400m)	253.78nm (470km)
34	Sikorsky Westland ASaC Mk 7	10,000ft (3,048m)	200nm (370km)
35	Tupolev Tu-126	35,100ft (10,700m)	174nm (322km)
36	Vickers Wellington Mk 1C	17,893ft (5,454m)	43.2nm (80km)
37	Vickers Wellington Mk XIV	16,404ft (5,000m)	43.2nm (80km)

Appendix II

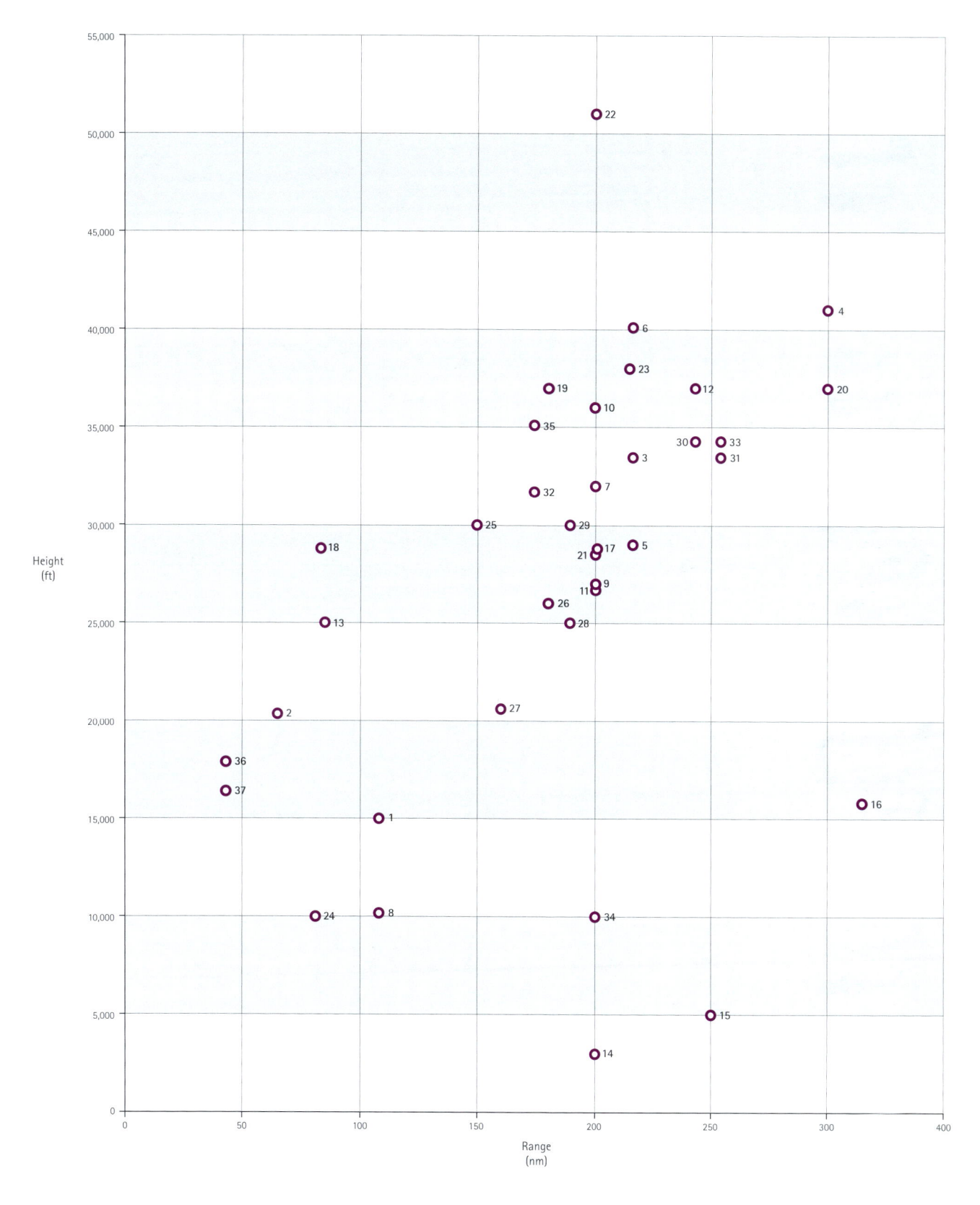

Beyond the Horizon – The History of Airborne Early Warning

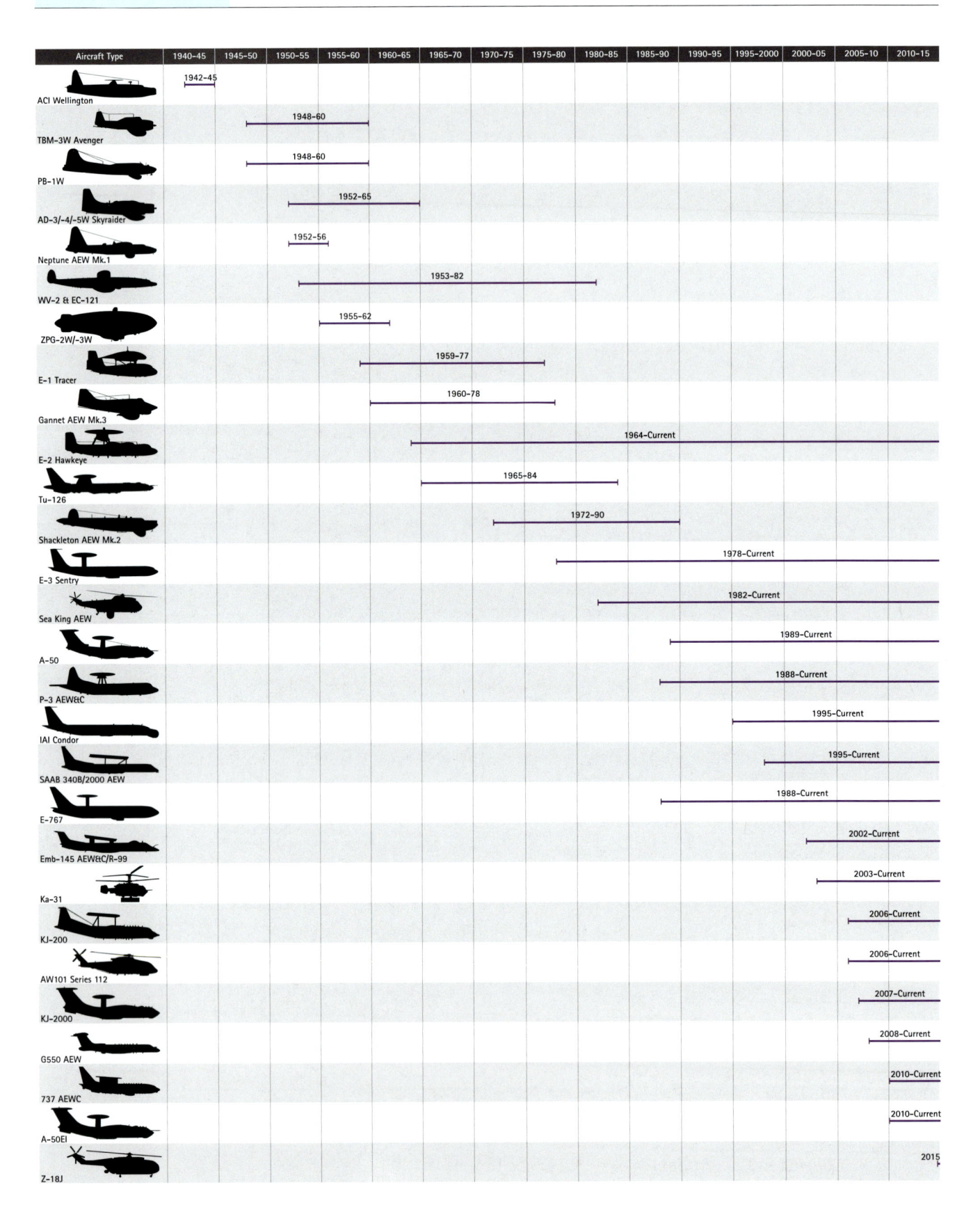

Aircraft Type	1940-45	1945-50	1950-55	1955-60	1960-65	1965-70	1970-75	1975-80	1980-85	1985-90	1990-95	1995-2000	2000-05	2005-10	2010-15
ACI Wellington	1942-45														
TBM-3W Avenger		1948-60	——————	——————											
PB-1W		1948-60	——————	——————											
AD-3/-4/-5W Skyraider			1952-65	——————	——————	——									
Neptune AEW Mk.1			1952-56	——											
WV-2 & EC-121				1953-82	——————	——————	——————	——————	——————	——					
ZPG-2W/-3W				1955-62	——										
E-1 Tracer					1959-77	——————	——————	—							
Gannet AEW Mk.3					1960-78	——————	——————	——							
E-2 Hawkeye					1964-Current	——————	——————	——————	——————	——————	——————	——————	——————	——————	——
Tu-126						1965-84	——————	——————	——						
Shackleton AEW Mk.2							1972-90	——————	——————	——					
E-3 Sentry								1978-Current	——————	——————	——————	——————	——————	——————	——
Sea King AEW									1982-Current	——————	——————	——————	——————	——————	——
A-50										1989-Current	——————	——————	——————	——————	——
P-3 AEW&C										1988-Current	——————	——————	——————	——————	——
IAI Condor											1995-Current	——————	——————	——————	——
SAAB 340B/2000 AEW											1995-Current	——————	——————	——————	——
E-767										1988-Current	——————	——————	——————	——————	——
Emb-145 AEW&C/R-99													2002-Current	——————	——
Ka-31													2003-Current	——————	——
KJ-200														2006-Current	——
AW101 Series 112														2006-Current	——
KJ-2000														2007-Current	——
G550 AEW														2008-Current	——
737 AEWC															2010-Current
A-50EI															2010-Current
Z-18J															2015

244

BIBLIOGRAPHY

ARMISTEAD, E. L., *Airborne Early Warning Systems: An International Guide to Technologies, Projects and Markets* (London: SMI Publishing, 2000)

ARMISTEAD, E. L., *AWACS and Hawkeyes – The Complete History of Airborne Early Warning Aircraft* (St Paul: MBI Publishing, 2002)   ISBN 978-0760311400

ASKINS, S., *Gannet from the Cockpit* (Ad Hoc Publications, 2008)   ISBN 978-0946958634

BABICH, V. K., *Russian Aviation and Air Defence Scientific-Technical Progress: Complex Combat Systems Yesterday, Today and Tomorrow* (in Russian) (Moscow: Drofa, 2004)   ISBN 978-5710799321

BARTON, D. K., AND LEONOV, S. A., *Radar Technology Encyclopedia* (Norwood: Artech House, 1998)   ISBN 978-0890068939

BOSLAUGH, D. L., *When Computers Went to Sea: The Digitization of the United States Navy* (Hoboken: Wiley IEEE, 2003)   ISBN 978-0471472209

BOWEN, E. G., *Radar Days* (Bristol: Institute of Physics Publishing, 1987)   ISBN 978-0852745908

BOWERS, P. M., *Boeing Aircraft Since 1916* (Annapolis: Naval Institute Press, 1988)   ISBN 978-3822896631

BROWN, L. A., *Radar History of World War II: Technical and Military Imperatives* (Boca Raton: CRC Press, 1999)   ISBN 0-7503-0659-9

CULL, B., NICOLLE, D., AND ALONI, S., *Wings over Suez* (London: Grub Street Publishing, 1996)   ISBN 978-1898697480

DARLING, K., *Lockheed Neptune* (Warpaint Books Ltd)

DAVIS, L., AND MENARD, D., *Douglas A-1 Skyraider* (Specialty Press, 1997)   ISBN 1 58007 066 3

DORR, R. F., *Douglas A-1 Skyraider* (Osprey Publishing Ltd, 1989)   ISBN 085-459060

Downs, E. (ed.), *Jane's Avionics 2003* (Coulsdon: Jane's Information Group, 2002) ISBN 978-0710624277

Duffy, P., and Kandalov, A., *Tupolev – The Man and His Aircraft* (Warrendale: Society of Automotive Engineers Inc, 1996)   ISBN 978-1560918998

Edgerton, D., *Warfare State: Britain 1920–1970* (Cambridge: Cambridge University Press, 2005)   ISBN 978-0521672313

Francillon, R. J., *Vietnam Air Wars* (Temple Press, 1987)   ISBN 0600-552845

Francillon, R. J., *Lockheed Aircraft since 1913* (New York: Brassey's Inc., 1988) ISBN 978-0851778051

Francillon, R. J., *Grumman Aircraft since 1929* (New York: Brassey's Inc., 1989) ISBN 978-0851778358

Geldhof, N., and Boerman, L., *TBM Avenger*, (Dutch Profile Publications)

Gibson, C., *Battle Flight* (Hikoki Publications, 2012)   ISBN 978-1902109268

Gibson, C., *The Air Staff & AEW* (Blue Envoy Press, 2013)   ISBN 978-0956195135

Gibson, C., *The Admiralty & AEW* (Blue Envoy Press, 2011)   ISBN 978-0956195128

Ginter, S., *Lockheed C-121 Constellation* (Naval Fighters Number 8) (Simi Valley: Steve Ginter Publishers, 1983)   ISBN 978-0942612080

Gordon, Y., Komissarov, D., *Soviet/Russian AWACS Aircraft* (Coulsdon: Midland Counties Publishing, 2005)   ISBN 978-1857802153

Gordon, Y., and Komissarov, D., *Chinese Aircraft: China's Aviation Industry since 1951* (Manchester: Hikoki Publications, 2008)   ISBN 978-1902109046

Hazell, S., *Warpaint Series No. 23 – Fairey Gannet* (Hall Park Limited, 2000) ISSN 1361-0369

Herwig, D., and Rode, H., *Luftwaffe Secret Projects: Ground Attack and Special Purpose Aircraft* (Hinckley: Midland Publishing, 2003)   ISBN 978-1857801507

Hirschel, E.-H., Prem, H., and Madelung, G., *Aeronautical Research in Germany: from Lilienthal until Today* (Berlin: Springer, 2004)   ISBN 978-3540406457

Hirst, M., *Airborne Early Warning Design, Development and Operations* (Osprey Publishing, 1983)   ISBN 0850455324

HOWARD, P. J., *Avro (Hawker Siddeley) Shackleton Marks 1 to 4* (Windsor: Profile Publications, 1972)   ISBN 0-853830223

HUGHES, K., AND DRANEM, W., *Douglas A-1 Skyraider – Warbird Tech Series Vol. 13* (North Branch: Specialty Press Publishers and Wholesalers, 1997) ISBN 978-0933424784

JACKSON, B. R., *Douglas Skyraider* (Fallbrook: Aero Publishers Inc., 1969) ISBN 9780816853038

JACKSON, P. (Ed.), *Jane's All The World's Aircraft 2004–2005* (Coulsdon: Jane's Information Group)   ISBN 978-0710626141

JONES, B., *Avro Shackleton* (Crowood Aviation Series) (Ramsbury: Crowood Press, 2002)   ISBN 978-1861264497

LACOMME, P., MARCHAIS, J-C., HARDANGE, J-P., AND NORMANT, E., *Air and Spaceborne Radar Systems: An Introduction* (Norwich: William Andrew Publishing, 2001) ISBN 978-1891121135

LAKE, D., *Growling Over The Oceans: The Royal Air Force Avro Shackleton, the Men, the Missions 1951–1991* (London: Souvenir Press Ltd, 2010) ISBN 978-0285638761

LAMBETH, B. S., *NATO Air War in Kosovo: A Strategic and Operational Assessment* (Santa Monica: RAND, 2000)   ISBN 978-0833032379

LAMBETH, B. S., *The Unseen War: Allied Air Power and the Takedown of Saddam Hussein* (Annapolis: Naval Institute Press, 2013)   ISBN 978-1612513119

LLOYD, A. T., *Boeing 707 & AWACS – In Detail & Scale Vol. 23.* (Shrewsbury: Aero Publishers, 1987)   ISBN 978-0830685332

MALONEY, S. M., *Enduring the Freedom: A Rogue Historian in Afghanistan* (Dulles: Potomac Books, 2007)   ISBN 978-1597970495

MICHEL III, M. L., *Clashes* (Naval Institute Press, 1997)   ISBN 1557505853

NEEDEL, A. A., *Science, Cold War and the American State: Lloyd V. Berkner and the Balance of Professional Ideals* (New York: Routledge, 2000)   ISBN 9789057026218

NICHOLS, J. B., AND TILLMAN, B., *On Yankee Station: The Naval Air War Over Vietnam* (New York: Naval Institute Press, 1987)   ISBN 978-1557504951

RAF Document, *No. 1453 Flight (AEW)*, (RAF Air 28 February 1955) Ref: 1453F/S.16/3/AIR

RAF Document, *RAF Operations Record Book No. 1453 Flt* (RAF Air August 1955) MOD Fm 540 No. 1453 Flt

RODRIGUES, L. F., *Larry's U.S. Navy Airship Picture Book* (Sacramento: EastWest Institute for Self-Understanding, 2004)   ISBN 978-0964086647

SANTANA, S., *Embraer EMB-145ISR – Program, Versions, Operators and Employment* (in Portuguese) (São Paulo: C&R Editorial, 2012)   ISBN 978-8599719169

SKOLNIK, M., *Radar Handbook* (Chicago: McGraw-Hill Professional, 2008) ISBN 978-0071485470

SMITH, R. G., *Guppy Pilot* (Western Oregon Web Press, 1998)

SMITH, P. C., *Douglas AD Skyraider* (Ramsbury: Crowood Press, 1998) ISBN 1-86126-249-3

SMITH, P. J. C., *Air Launched Doodlebugs: The Forgotten Campaign* (Barnsley: Pen and Sword Aviation, 2006)   ISBN 978-1844154012

SPACEK, J., SPURNY, J., AND MARTINEC, J., *Sea King in Detail – Westland Sea King and its Export Variants* (Prague: Wings & Wheels Publications, 2006) ISBN 80-86416-54-2

STAFRACE, C., *Grumman S2F Tracker, TF-1 Trader and WF-2 Tracer* (Warpaint Books Ltd)

STREETLY, M., *Jane's Radar and Electronic Warfare Systems 2002* (Coulsdon: Jane's Information Group, 2002)   ISBN 978-0710624451

SWANBOROUGH, G., AND BOWERS, P. M., *United States Navy Aircraft since 1911* (New York: U.S. Naval Institute Press, 1990)   ISBN 978-0870217920

TANNER, R. M. M.B.E., *History of Air To Air Refuelling* (Pen & Sword Aviation, 2006) ISBN 1844152723

THOMPSON, S. A., *B-17 in Blue* (Aero Vintage Books, 1993)   ISBN 0963754335

US Navy Publication, AN 01-40ALB-1 *(Pilot's Handbook – Navy Models AD-3W/ AD-4W)*   No ISBN

US Navy Publication, AN 01-40ALB-1, 1 November 1955, F*LIGHT Handbook Interim Revision No.1 (Navy Model AD-3W -4W Aircraft)* pp65-95

US Navy Publication, NAVWEPS 01-195PDA-501-15 April 1960, revised 15 July 1960, FLIGHT HANDBOOK NAVY MODEL ZPG-3W

US Navy Publication, NAVAIR 00-11AW-1, *Standard Aircraft Characteristics – Navy Model E-1B Aircraft*

US Navy Publication, NAVAIR 01-85WBA-1 *(NATOPS FLIGHT MANUAL NAVY MODEL TE-2A, E-2B)*

Various, *A History of the Douglas Skyraider AEW.1* (British Aviation Research Group, 1974)   ISBN 0906339014

VELEK, M., OVCÁCÍK, M. AND SUSA, K., *Fairey Gannet: Antisubmarine and Strike Variants AS Mk.1, AS Mk.4* (Prague: Mark I Ltd, 2007)   ISBN 978-8086637044

WILLIAM, G. K., 'AWACS and JSTARS', in *Technology and the Air Force: A Retrospective Assessment*, NEUFELD, J., WATSON, G. M. JR., AND CHENOWETH D., (eds) (Washington, D.C.: Diane Publishing, 1997)   ISBN 978-1410201850

VERBA, V. S., *Early Warning and Control Aviation Systems – Status and Trends of Development* (in Russian) (Moscow: Radiotekhnika, 2008)   ISBN 978-5880702046

WINCHESTER, J., *Boeing 707/720* (Airlife's Classic Airliners) (Shrewsbury: Airlife Publishing, 2002)   ISBN 978-1840373110

WOOLEY, P. J., *Japan's Navy: 1971–2000, Politics and Paradox* (Boulder: Lynne Rienner Publishers, 2000)   ISBN 1-55587-819-9

YOUNGERS, C., AND ROSIN, E., *Drugs and Democracy in Latin America: the Impact of U.S. Policy.* (Boulder: Lynne Rienner Publishers, 2005)   ISBN 1-58826-278-2

Various volumes of *Aerospaço* (the official publication of the Air Control Department, Brazilian Air Force), from 2009 to 2013; *Air Fleet* magazine, from 2003 to 2008; *Air Force* (Magazine of The Air Force Association of Canada), 2003; *Air Forces Monthly* magazine, from 2000 to 2012; *Air World* magazine, 1998; *Aviation Industries Journal* (Iran), 2008; *Aviatsiya i Vremia* magazine, 2003; *Combat Aircraft Monthly* magazine, from 2010 to 2012; *Defensa* magazine, 2006; *Eyes of the Eagle* (the official publication of the 552nd Air Control Wing, USAF); *Flight International* magazine, from 1975 to 2004; *Flypast*, 1987; *Global Aviator* magazine (South Africa), 2013; *International Journal of Electrical Engineering Education*, 1999; *Lincoln Laboratory Journal* (United States), 2000; *Naval Aviation News* (the official publication of the US Navy), from 1959 to 1960; *O Guardião* (official publication of the 2º Esquadrão of 6º Grupo de Aviação, Brazilian Air Force), from 2006 to 2009; *People's Liberation Army Air Force Academic Review Monthly* (China), 2006; *Vzlet* magazine, from 2009 to 2013; *Wings of Gold* (the official publication of the US Navy), from 2000 to 2005, and interviews with various Brazilian, British, Canadian, German, Greek and North American officers, pilots and ground personnel, and personal notes of both the authors.

INDEX

Aircraft

AgustaWestland AW101 Series 112 159, 186, 242, 244
Airbus C295 AEW 172
Arado Ar 234 82
Avro Shackleton AEW.Mk 2 36, 73, 74–76, 90, 92, 101, 188, 242, 244
BAe Nimrod AEW.Mk 3 8, 74–81, 92, 119, 148, 151, 164, 177
Beriev A-50 98, 148, 155, 190, 242, 244
Beriev/IAI A-50EI 190, 191, 242, 242
Beriev A-50U 99, 166, 167, 190, 191
Beriev A-100 167
Boeing 737 AEW&C 162, 166, 171, 175, 192, 242, 244
Boeing E-3/ E-3A Sentry 79–81, 83–87, 89, 90, 95–97, 118–120, 124, 134, 135, 137, 142, 159, 164, 165, 179, 192, 194, 242, 244
Boeing E-3B Sentry 96, 137, 179, 180, 194, 195
Boeing E-3C Sentry 124, 129, 130, 161, 179–181, 195
Boeing E-3D Sentry 8, 101, 102, 118, 120–124, 128, 130, 131, 135–138, 141, 142, 177, 194, 195
Boeing E-3F Sentry 120, 124, 137, 138, 141, 152, 153, 194
Boeing E-3G Sentry (Block 40/45) 181
Boeing E-7A Wedgetail (Australia) 143, 144
Boeing E-767 160–162, 196, 197, 242, 244
Boeing EC-137D 81, 82
Boeing PB-1W 23, 198, 242
Changhe Z-18J 151, 152, 244
Douglas AD-3W Skyraider 25, 26, 46, 202, 203, 216, 242
Douglas AD-4W Skyraider 26, 27, 202, 203, 242
Douglas AD-5W Skyraider 27, 28, 71, 202, 203, 242
Douglas Skyraider AEW.Mk 1 26, 27, 36, 203, 242

Embraer 145 AEW&C 146, 204, 242
Embraer 145H 137, 139, 140, 142, 154, 204, 205
Embraer EMB-145I 142, 145, 146, 156
Embraer E-99 145, 146, 154, 204, 205, 242
Fairey Gannet AEW.Mk 3 8, 27, 37–40, 63, 74, 188, 206
Goodyear ZPG-2W/-3W 31, 32, 208, 209, 242, 244
Grumman AF-2W Guardian 31
Grumman E-1B Tracer 8, 40, 41, 47, 63, 65–67, 71–73, 81, 242, 244
Grumman E-2 Hawkeye 40, 212, 244
Grumman E-2A Hawkeye 42, 52, 53, 66, 212–214, 242
Grumman E-2B Hawkeye 52, 53, 65, 66, 71–73, 174, 212, 213, 242
Grumman E-2C Hawkeye 72, 73, 87, 89, 91, 93–95, 108, 110, 112–114, 128, 130, 134, 137, 138, 152, 153, 157, 158, 160, 162, 163, 170, 174, 182, 212–214, 242
Grumman E-2C Basic 212–214
Grumman E-2C Group 0 87, 152, 157, 160, 213, 214
Grumman E-2C Group I 212–214
Grumman E-2C Group II 152, 174, 182, 183, 214
Grumman E-2C Group II Plus/Group III/ Hawkeye 2000 153, 182, 183, 212
Grumman E-2D Advanced Hawkeye 95, 153, 162, 183, 184, 212–214, 242
Grumman TBM-3W/W2 Avenger 22, 25, 216, 242, 244
Grumman WF-2 Tracer (see Grumman E-1 Tracer)
Gulfstream G 550 CAEW Nachshon Eitam 157–159, 170, 171, 218, 242, 244
Hawker Siddeley HS748AEW 155
IAI/ Boeing Condór/ Phalcon 146–148, 157, 158, 220, 242, 244
Ilyushin Il-76 Adnan 104–106, 190, 191

Ilyushin Il-76 Baghdad 104, 105, 190, 191
Kamov Ka-31 151, 152, 157, 168, 222, 242, 244
Lockheed C-130AEW 153
Lockheed EC-121 Warning Star 29–31, 43, 46, 55, 57–59, 62, 63, 65, 229, 242, 244
Lockheed EC-121D/RC-121D 29, 30, 48, 53–56, 59, 60
Lockheed EC-121H 31
Lockheed EC-121K 61
Lockheed EC-121Q 30
Lockheed EC-121T 30, 64
Lockheed NC-130H 183, 184
Lockheed Neptune AEW.Mk 1 34, 35, 224, 242, 244
Lockheed P2V-5 Neptune 31, 33, 34
Lockheed PO-2W (see Lockheed WV-2)
Lockheed WV-1 24, 28, 229
Lockheed WV-2 8, 28–30, 49–51, 62, 67, 228, 229, 242, 244
Lockheed WV-3 229
Lockheed Martin P-3 AEW 114, 115, 226, 242, 244
Saab 100S 154, 173, 242
Saab 340 AEW&C 154, 175, 176, 230, 231, 244
Saab 2000 AEW&C 142, 165, 169, 230, 231, 244
Shaanxi Y-8AEW (ZDK-03) 150, 242
Shaanxi Y-8J 151, 242
Shaanxi KJ-200/Y-8W 149, 150, 232, 242, 244
Shaanxi KJ-500 150, 151
Sikorsky HR2S-1W 31
Sikorsky SH-3 AEW Sea King 171, 172 238, 244
Tupolev Tu-4 (KJ-1) 147,148
Tupolev Tu-126 44–46, 63, 96, 97, 191, 234, 242
Vickers Wellington 16–18, 236, 242
Westland Sea King AEW. Mk 2 92, 93, 118, 238, 242

251

Westland Sea King ASaC.Mk 7 8, 132, 133, 136, 142, 178, 179, 238, 239, 242
Xi'an KJ-2000 149–151, 190
Xi'an JZY-01 151, 152

Australia
Royal Australian Air Force 114, 143
RAAF No. 2 Squadron 143, 144

Brazil
Brazilian Air Force 145
2º/6º Grupo de Aviação 'Esquadrão Guardião' 145

Chile
Chilean Air Force 146
Grupo de Aviación No. 10 147

China
4th Air Regiment/2nd Naval Aviation Division 150
76th Electronic Warfare Regiment, 26th Special Mission Division 149
People's Liberation Army Air Force (PLAAF) 147
People's Liberation Army Navy (PLAN) 151

Conflicts and Operations
Chechnya 99
Cuban Missile Crisis 43
Iran-Iraq War 89
Korean War 20, 22, 26
Suez Crisis 25, 26
Operation Active Endeavour 133
Operation Afghan Assist 134
Operation Allied Force 121, 124
Operation Big Eye 53, 54, 56
Operation Cast Lead 158
Operation College Eye 53, 54, 58, 60, 61
Operation Deference 135
Operation Deliberate Force 120, 121
Operation Deny Flight 118, 120, 121
Operation Desert Shield 102, 104, 106, 108
Operation Desert Storm 8, 99, 106–108
Operation Eagle Assist 127
Operation Ellamy 136
Operation Enduring Freedom 129
Operation European Liaison Force One (ELF-1) 89
Operation Frequent Wind 71
Operation Héracles 129
Operation Iraqi Freedom 130
Operation Just Cause 96
Operation Linebacker 65
Operation Linebacker II 67
Operation Maritime Monitor 118
Operation Mole Cricket 19 94
Operation New Dawn 133
Operation Northern Watch 131
Operation Orchard 158
Operation Proven Force 108
Operation Provide Comfort 111
Operation Rolling Thunder 57, 63, 65
Operation Serval 141
Operation Sky Monitor 118
Operation Southern Watch 111, 131
Operation Telic (see Operation Iraqi Freedom)
Operation Thunder 112
Operation Unified Protector 135–137, 140, 141
Operation Urgent Fury 95
Operation Vapour 8, 18
Operation Veritas (see Operation Enduring Freedom)
Vietnam War 8, 43, 47, 59, 63, 71, 93
World War II (WWII) 6, 7, 15, 26, 91

Egypt
Egyptian Air Force 152
87 Squadron, 601 AEW Brigade 152

France
Escadron de Détection et de Contrôle Aéroportés (ECDA) 36 'Berry' 153
Flottille 4F 128, 137, 153
French Air Force/Armée de l'Air 120, 124, 137, 152, 153, 177
French Navy 24–26, 137, 138, 153

Greece
380 MASEPE 137, 139; 140
Hellenic Air Force 8, 138, 139, 153, 154

India
Defence Research and Development Organisation (DRDO) 156
INAS 339 'Falcons' (Indian Navy) 157
No. 50 Squadron (Air Force) 155

Iran
73rd Tactical Transport Squadron 105
Imperial Iranian Air Force (IIAF) 86
Islamic Republic of Iran Air Force (IRIAF) 89, 105
Simorgh 105

Iraq
Iraqi Air Force 11, 104
Adnan, Adnan-2 104, 105
Baghdad-1, Baghdad-2 104, 105

Israel
Israel Aircraft Industries 146–148, 157, 158, 170–172, 179, 220, 221
Israel Defense Forces/Air Force (IDF/AF) 11, 87
No. 122 Nachshon Squadron 95
No. 192 Hawkeye Squadron 87

Italy
3° Gruppo Elicotteri 159

Japan
601 Hikotai 160
602 Hikotai 161
603 Hikotai 160
Japan Air Self Defense Force 160
Japan Maritime Self Defense Force 25

Mexico
Escuadron de Vigilancia Aerea 163
Primer Escuadron Alerta Temprana y Reconocimiento 163

NATO
NATO Airborne Early Warning & Control Force (NAEWF) 90, 118, 134, 142, 164

Pakistan
No. 4 Squadron 166

Radar
AN/APS-14 19
AN/APS-20 19, 20, 21, 25, 28, 30–33, 46, 51
AN/APS-20A 198
AN/APS-20B/E 49, 81
AN/APS-20F 37, 38, 74, 78, 207
AN/APS-20F (I) 40, 188, 207
AN/APS-42 30, 229
AN/APS-45 28, 30, 50, 51, 53, 229
AN/APS-62 209
AN/APS-69 31, 32, 209
AN/APS-82 40, 81, 211
AN/APS-95 30, 32, 53, 209, 229
AN/APS-96 214
AN/APS-103 30, 57, 229
AN/APS-120 73, 214
AN/APS-125 87, 112, 214
AN/APS-138 87, 114, 138, 174, 214

AN/APS-139 214
AN/APS-145 115, 144, 152, 174, 214
ASV Mk III 17, 18
AN/APY-1 81, 82, 85, 97, 179, 195
AN/APY-2 161, 164, 179, 197
AN/APY-9 183, 215
E-801 Oko 168, 223
EL/M-2075 Phalcon 191
EL/M-2085 218
EL/M-2090 155
Ericsson PS-890 Erieye 145, 163, 205, 231
Liana 44, 45, 191, 235
MM/APS-784 186
Northrop Grumman MESA D-Band 193
SCR-720 19
Shmel 97, 98, 167, 191
Shmel-2 191
Thorn-EMI Searchwater/Searchwater 2000 AEW 92, 151, 171, 178, 179, 239
Vega Premier 167

Russia
67th Independent Airborne Early Warning Squadron 45
859th Evaluation Centre 168
Ivanovo air base 166
Russian Air Force 99, 100, 148, 166, 167

Soviet Union
Soviet Union/USSR 28, 29, 43, 46, 63, 64, 75, 99, 165, 235

Saudi Arabia
No. 18 Squadron 102, 169
Royal Saudi Air Force (RSAF) 12, 102, 104, 108, 169, 194, 231

Singapore
111st Squadron 170
Republic of Singapore Air Force 170

South Korea
Republic of Korea Air Force (RoKAF) 171
271st AEW&C Flight Squadron 171

Spain
Spanish Navy 171, 172

Sweden
72. Ledningsflygdivisionen 174
ASC 890/FSR 890 173, 174
Swedish Air Force 172–174

Taiwan
Republic of China Air Force (RoCAF) 174
2nd Early Warning Squadron 174

Thailand
No. 702 Squadron 175
Royal Thai Air Force 175

The Netherlands
Royal Netherlands Navy 24, 25

Turkey
131 Filo 176
Turkish Air Force 13, 175, 176

United Arab Emirates
United Arab Emirates Air Force 176

United Kingdom
Coastal Command 15, 17, 18, 33, 75, 76
de Havilland Comet 4C 79
Fleet Air Arm 11, 26, 63, 74
Royal Air Force (RAF) 6, 8, 12, 15, 74, 122, 128, 177, 247
Royal Air Force Squadrons
 No. 8 Squadron 74, 75, 77, 90, 101, 118, 121, 122, 128, 136
 No. 23 Squadron 121–123
 No. 1453 Flight 33–35, 247
Royal Navy (RN) 8, 12, 26, 27, 33, 36, 37, 73, 116, 119, 132, 142, 178
Royal Navy Squadrons
 778 NAS 26
 824 NAS 92
 849 NAS 26, 27, 37, 38, 46, 74, 92, 118, 120, 132, 133, 178
Royal Radar Establishment (RRE) 12, 15
Vanguard Flight 33

United States
Airborne Warning and Control Wing (AWACW) 86, 89, 102
EC-121 Rivet Gym 61, 62, 64
EC-121 Rivet Top 61, 62, 64
EC-121 Quick Look 58, 59
Distant Early Warning Line (DEW Line) 28–30
NORAD 28, 31, 53, 85, 86, 128
Project Cadillac and Cadillac 2 18, 20, 21, 32
USAF 13, 89, 90, 95, 102–104, 106–108, 111, 112, 114, 115, 118, 119, 123, 124, 127–131, 137, 138, 141, 153, 159, 164, 165, 177, 179, 180, 181, 194, 249

U.S. Air Force Squadrons
 551st Airborne Early Warning and Control Wing 31, 53
 552nd Airborne Early Warning and Control Wing (AEWCW) 53, 54, 95, 96, 104
 552nd Air Control Wing (ACW) 54, 111
 960th AWACSS 82, 86
 961st AWACSS 82, 86
 963rd AWACS 82, 86
 964th AWACS 84, 86
 965th AWACS 84, 86
 966th AWACS 84, 86
 4701st Airborne Early Warning and Control Squadron 30
 7740th Provisional Wing 108
U.S. Coast Guard (USCG) 112–114, 116
U.S. Customs Border Protection (CBP) 114–117
U.S. Marine Corps Squadrons
 VMA-121 27
U.S. Navy Squadrons
 RVAW-110 42, 52, 71
 RVAW-120 41
 Task Force (TF-77) 48, 49, 58
 VAW-1 20, 25, 26, 28, 40, 41
 VAW-2 20
 VAW-3 25
 VAW-11 43
 VAW-12 40
 VAW-77 114
 VAW-110 66
 VAW-111 65, 66
 VAW-112 128
 VAW-113 109, 128
 VAW-114 53, 65
 VAW-115 65, 110, 170
 VAW-116 110
 VAW-117 128, 182, 183
 VAW-120 184
 VAW-121 109, 128, 134
 VAW-122 95, 96
 VAW-123 73, 110, 128
 VAW-124 72, 91, 110, 128
 VAW-125 110, 128, 184
 VAW-126 91, 110, 182
 VC-11 22, 26
 VW-1 22–24, 28, 29, 49–52
 ZW-1 32

ACIG

Online since 1999
ACIG is a multi-national project
dedicated to research about
air wars and air forces since 1945

Associated authors, photographers, artists and contributors
have published 32 books, hundreds of articles and artworks.
Multiple research projects are going on and we are
looking forward for your contributions:
join us at ACIG.info forum!

www.acig.info

---

www.aviationgraphic.com

**We are on FACEBOOK
Join our community!**

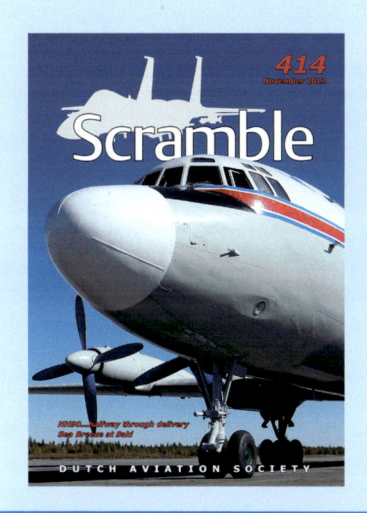

Read all the latest Civil & Military aviation news in Scramble Magazine

Also for Smartphone & Tablet

starting at only $2,75 a month

Visit www.scramble.nl for air force overviews, serial rundowns and more

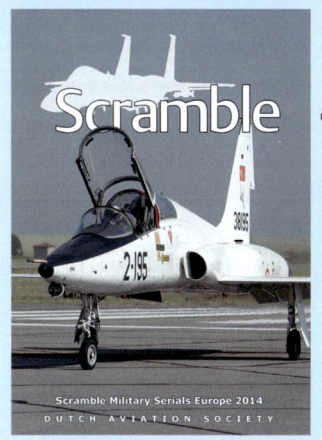

Military Serials Europe 2014
Included in the 2014 edition are European (except Russia) and permanently stationed Singapore and United States military aircraft in Europe. For every country, all types are listed that have at least one aircraft of that type active as of 1 February 2014, not including those only used by technical schools

Military Serials North America 2014
After seven years of absence, we are pleased to introduce a new edition of the SMS North America. This 223 pages book includes an Order of Battle containing a rundown on where all USAF, US Army, US Navy, USMC, NASA & Coastguard aircraft and helicopters are based

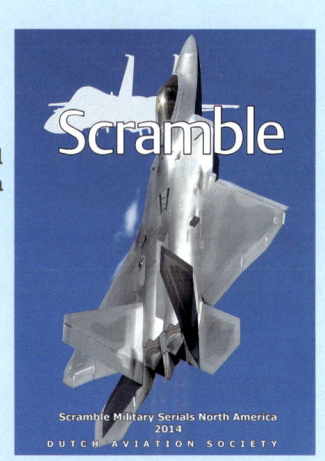

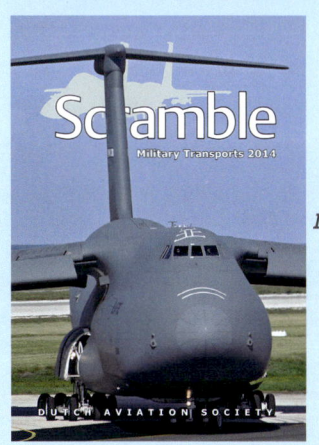

Military Transports 2014
New in our inventory is the Scramble Military Transports 2014. The book covers all active transport, tanker, maritime patrol, liaison and training aircraft of the world's armed forces, as well as government operated aircraft

The 'must have' during your trips:
The Scramble database app
With unlimited database searches & saves

Available on the App Store
ANDROID APP ON Google play

Visit: www.scramble.nl/shop

DUTCH AVIATION SOCIETY

HARPIA PUBLISHING

Glide With Us Into The World of Aviation Literature

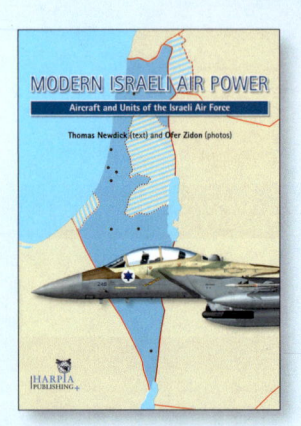

Modern Israeli Air Power
Aircraft and Units of the Israeli Air Force
Thomas Newdick (text) and Ofer Zidon (photos)
256 pages, 28 x 21 cm, softcover
35.95 Euro, 978-0-9854554-2-2

Israel remains the cornerstone of Middle East conflicts and tensions, and the spearhead of Israeli military might remains the Air Corps (Kheil Ha'Avir) of the Israeli Defence Forces. Renowned for its continuous efforts to maintain dominance in every dimension of air warfare, improve its capabilities, and outsmart its opponents, the Israeli Air and Space Force has recently been moving away from preparations for interstate wars towards improving its potential to wage asymmetric conflicts, counter-insurgency campaigns and special operations.

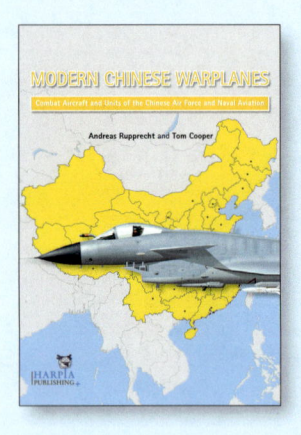

Modern Chinese Warplanes
Combat Aircraft and Units of the Chinese Air Force and Naval Aviation
Andreas Rupprecht and Tom Cooper
256 pages, 28 x 21 cm, softcover
35.95 Euro, ISBN 978-0-9854554-0-8

Much of the fascination that Chinese military aviation holds for the analyst and enthusiast stems from the thick veil of secrecy that surrounds it. This uniquely compact yet comprehensive directory serves as a magnificently illustrated, in-depth analysis and directory of modern Chinese air power. It is organised in three parts: the most important military aircraft and their weapons found in Chinese service today; aircraft markings and serial number systems; and orders of battle for the People's Liberation Army Air Force and Naval Air Force.

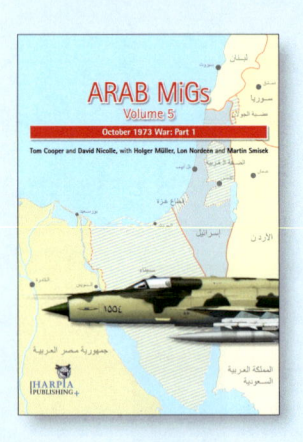

Arab MiGs Volume 5 October 1973 War: Part 1
Tom Cooper and David Nicolle, with Holger Müller, Lon Nordeen and Martin Smisek
256 pages, 28 x 21 cm, softcover
35.95 Euro, ISBN 978-0-9854554-4-6

On 6 October 1973, the Egyptian and Syrian air arms launched an attack on Israeli military installations on the Sinai Peninsula and in the Golan Heights. For Israel, it was a war that started with a surprise and alarming losses in men and material, and was characterised by the deployment of advanced weapons and equipment. For the Arabs it was a war of revenge, in the best traditions of 'guts and glory'. *Arab MiGs Volume 5* provides a detailed record of aerial warfare during the opening phases of the October 1973 conflict. While concentrating on the Arab experiences, *Arab MiGs Volume 5* is the first comprehensive analysis of the aerial operations waged by both sides in this conflict.

THE AVIATION BOOKS OF A DIFFERENT KIND
UNIQUE TOPICS | IN-DEPTH RESEARCH | RARE PICTURES | HIGH PRINTING QUALITY

www.harpia-publishing.com